Parallel Comput
in
Quantum Chemistry

Parallel Computing *in* Quantum Chemistry

Curtis L. Janssen

Ida M. B. Nielsen

CRC Press is an imprint of the
Taylor & Francis Group, an **informa** business

This manuscript has been authored by Sandia National Laboratories, Contract No. DE-AC04-94AL85000 with the U.S. Department of Energy and Limit Point Systems, Inc. subcontract LF-5537 with Sandia. The publisher, by accepting the manuscript for publication, acknowledges that the United States Government retains a non-exclusive, paid-up, irrevocable, world-wide license to publish or reproduce the published form of this manuscript, or allow others to do so, for United States Government purposes.

CRC Press
Taylor & Francis Group
6000 Broken Sound Parkway NW, Suite 300
Boca Raton, FL 33487-2742

First issued in paperback 2019

ISBN-13: 978-1-4200-5164-3 (hbk)
ISBN-13: 978-0-367-38750-1 (pbk)

Library of Congress Cataloging-in-Publication Data

Janssen, Curtis L.
Parallel computing in quantum chemistry / Curtis L. Janssen and Ida M.B. Nielsen.
p. cm.
Includes bibliographical references and index.
ISBN 978-1-4200-5164-3 (acid-free paper)
1. Quantum chemistry--Data processing. 2. Parallel programming (Computer science) 3. Parallel processing (Electronic computers) I. Nielsen, Ida M. B. II. Title.

QD462.6.D38J36 2008
541'.280285435--dc22 2007048052

Visit the Taylor & Francis Web site at
http://www.taylorandfrancis.com

and the CRC Press Web site at
http://www.crcpress.com

Contents

Preface

This book is intended to serve as a reference for the design and implementation of parallel quantum chemistry programs. Development of efficient quantum chemistry software capable of utilizing large-scale parallel computers requires a grasp of many issues pertaining to both parallel computing hardware and parallel programming practices, as well as an understanding of the methods to be implemented. The text provides an in-depth view of parallel programming challenges from the perspective of a quantum chemist, including parallel computer architectures, message-passing, multi-threading, parallel program design and performance analysis, as well as parallel implementation of important electronic structure procedures and methods such as two-electron integral computation, Hartree–Fock and second-order Møller–Plesset perturbation (MP2) theory, and the local correlation method LMP2. Some topics relevant to parallel computing in quantum chemistry have not been included in this book. Thus, performance tools and debuggers are not treated, parallel I/O is only briefly discussed, and advanced electronic structure methods such as coupled-cluster theory and configuration interaction are not covered.

We will assume that the reader has a basic understanding of quantum chemistry, including Hartree–Fock theory and correlated electronic structure methods such as Møller–Plesset perturbation theory. Readers can find introductory discussions of these methods in, for example, Jensen[1] and the classic Szabo and Ostlund text.[2] A comprehensive and somewhat more advanced treatment of electronic structure theory can be found in Helgaker, Jørgensen, and Olsen.[3] No prior experience with parallel computing is required, but the reader should be familiar with computer programming and programming languages at the advanced undergraduate level. The program examples in the book are written in the C programming language, and at least a rudimentary knowledge of C will therefore be helpful. The text by Kernighan and Ritchie[4] covers all C features used in this book.

Scope and Organization of the Text

This book is divided into two parts. In Part I we will discuss parallel computer architectures as well as parallel computing concepts and terminology with a focus on good parallel program design and performance analysis. Part II contains detailed discussions and performance analyses of parallel algorithms for a number of important and widely used quantum chemistry procedures and methods.

An outline of the contents of each chapter is given below.

Chapter 1: Introduction

Here we provide a brief history of parallel computing in quantum chemistry and discuss trends in hardware as well as trends in the methods and algorithms of quantum chemistry. The impact of these trends on future quantum chemistry programs will be considered.

Chapter 2: Parallel Computer Architectures

This chapter provides an overview of parallel computer architectures, including the traditional Flynn classification scheme and a discussion of computation nodes and the networks connecting them. We also present an overall system view of a parallel computer, describing the hierarchical nature of parallel architecture, machine reliability, and the distinction between commodity and custom computers.

Chapter 3: Communication via Message-Passing

This chapter covers message-passing, one of the primary software tools required to develop parallel quantum chemistry programs for distributed memory parallel computers. Point-to-point, collective, and one-sided varieties of message-passing are also discussed.

Chapter 4: Multi-Threading

The importance of multi-threading will continue to increase due to the emergence of multicore chips. Parallelization by means of multi-threading is discussed as well as hybrid multi-threading/message-passing approaches for utilizing large-scale parallel computers.

Chapter 5: Parallel Performance Evaluation

Design and implementation of efficient parallel algorithms requires careful analysis and evaluation of their performance. This chapter introduces idealized machine models along with measures for predicting and assessing the performance of parallel algorithms.

Chapter 6: Parallel Program Design

This chapter discusses fundamental issues involved in designing and implementing parallel programs, including the distribution of tasks and data as well as schemes for interprocessor communication.

Chapter 7: Two-Electron Integral Evaluation

An important, basic step performed in most quantum chemistry programs is the computation of the two-electron integrals. Schemes for parallel computation of these integrals and detailed performance models incorporating load imbalance are discussed.

Chapter 8: The Hartree–Fock Method

The Hartree–Fock method is central to quantum chemistry, and an efficient Hartree–Fock program is an essential part of a quantum chemistry program package. We outline the Hartree–Fock procedure and present and analyze both replicated data and distributed data Fock matrix formation algorithms.

Chapter 9: Second-Order Møller–Plesset Perturbation Theory

Second-order Møller–Plesset (MP2) perturbation theory is a widely used quantum chemical method for incorporating electron correlation. This chapter considers parallel computation of MP2 energies, comparing the performance achievable with simple and more sophisticated parallelization strategies.

Chapter 10: Local Møller–Plesset Perturbation Theory

Local correlation methods represent an important new class of correlated electronic structure methods that aim at computing molecular properties with the same accuracy as their conventional counterparts but at a significantly lower computational cost. We discuss the challenges of parallelizing local correlation methods in the context of local second-order Møller–Plesset perturbation theory, illustrating a parallel implementation and presenting benchmarks as well.

Appendix A: A Brief Introduction to MPI

The Message-Passing Interface (MPI) is the primary mechanism used for explicit message-passing in scientific computing applications. This appendix briefly discusses some of the most commonly used MPI routines.

Appendix B: Pthreads

Pthreads is a standard for creating and managing multiple threads. We give a brief introduction to multi-threaded programming with Pthreads, including an example Pthreads program.

Appendix C: OpenMP

OpenMP is a set of compiler extensions to facilitate development of multi-threaded programs. We describe these compiler extensions, using example source code illustrating parallel programming with OpenMP.

References

1. Jensen, F. *Introduction to Computational Chemistry*. Chichester, UK: John Wiley & Sons, 1999.
2. Szabo, A., and N. S. Ostlund. *Modern Quantum Chemistry*, 1st revised edition. New York: McGraw-Hill, 1989.
3. Helgaker, T., P. Jørgensen, and J. Olsen. *Molecular Electronic-Structure Theory*. Chichester, UK: John Wiley & Sons, 2000.
4. Kernighan, B. W., and D. M. Ritchie. *The C Programming Language*, 2nd edition. Englewood Cliffs, NJ: Prentice Hall, 1988.

Acknowledgments

We are indebted to many people who provided support and advice in the preparation of this book. Lance Wobus of Taylor & Francis encouraged us to undertake the writing of the book, and Jennifer Smith expertly guided us through the final stages of manuscript preparation. We thank Craig Smith and Kurt Olsen at Sandia National Laboratories for facilitating the project, and we appreciate the assistance of Gregg Andreski, who prepared most of the artwork. Jan Linderberg kindly provided historical references, and Shawn Brown, Joe Kenny, and Matt Leininger made insightful comments on various drafts of the manuscript. Ron Minnich, Helgi Adalsteinsson, Jim Schutt, Ed Valeev, Daniel Crawford, Theresa Windus, and David Bernholdt provided stimulating discussions, feedback, and technical data. Finally, CLJ would like to thank Mike Colvin for introducing him to parallel computing.

Funding Statement

Sandia is a multiprogram laboratory operated by Sandia Corporation, a Lockheed Martin Company, for the United States Department of Energy's National Nuclear Security Administration under contract DE-AC04-94AL85000.

Authors

Curtis L. Janssen is a Distinguished Member of the Technical Staff at Sandia National Laboratories and holds a Ph.D. in theoretical chemistry from the University of California at Berkeley. He has an extensive publication record in the areas of quantum chemistry methods development and high-performance computing, and he is the lead developer of the MPQC program suite.

Curtis L. Janssen
Sandia National Laboratories
7011 East Ave.
Livermore, CA 94550
cljanss@sandia.gov

Ida M. B. Nielsen is a scientist at Sandia National Laboratories and holds a Ph.D. in theoretical chemistry from Stanford University. She has published numerous articles on the development and application of quantum chemical methods and high-performance computing and is one of the core developers of the MPQC program suite.

Ida M. B. Nielsen
Sandia National Laboratories
7011 East Ave.
Livermore, CA 94550
ibniels@sandia.gov

Disclaimer of Liability

Trademarks

Part I

Parallel Computing Concepts and Terminology

1

Introduction

The need for parallel software for scientific computing is ever increasing. Supercomputers are not only being built with more processors, but parallel computers are also no longer limited to large machines owned and managed by high-performance computing centers; parallel desktop computers are increasingly widespread, and even laptop computers have multiple processors. Development of scientific computing software must adapt to these conditions as parallel computation becomes the norm rather than the exception. In the field of quantum chemistry, additional factors contribute to the need for parallelism. Quantum chemistry has become an indispensable tool for investigating chemical phenomena, and quantum chemical methods are employed widely in research across many chemical disciplines; this widespread use of quantum chemistry reinforces the importance of rapid turnaround computations, which can be addressed by parallel computing. Additionally, for quantum chemistry to continue to be an integral part of chemical research, quantum chemical methods must be applicable to the chemical systems of interest, including larger molecules, and parallelism can play an important role in extending the range of these methods. Parallel implementations can broaden the scope of conventional quantum chemical methods, whose computational cost scales as a high-degree polynomial in the molecular size, and enable the treatment of very large molecular systems with linear-scaling or reduced-scaling methods.

In the following, we will first give a brief historical sketch and current perspective of parallel computing in quantum chemistry. We will then discuss trends in hardware development for single-processor and parallel computers as well as trends in parallel software development, including the parallel programming challenges following the emergence of new quantum chemical methods and the changes in hardware.

1.1 Parallel Computing in Quantum Chemistry: Past and Present

Although the history of parallel computing in quantum chemistry is closely linked to the development of parallel computers, the concept of parallel computing was utilized in quantum chemistry even before the introduction of electronic computers into the field. In the late 1940s and early 1950s, intrigued by the idea of parallel computations, Per-Olov Löwdin in Uppsala organized a group of graduate students to carry out numerical integrations using FACIT Electric desk calculators. In six months, this "parallel computer" evaluated more than 10,000 multicenter integrals, employing Simpson's rule with a correction term. These integrals allowed the computation of wave functions that made it possible to explain the cohesive and elastic properties as well as the behavior under very high pressure of a large number of alkali halides.[1,2]

The use of parallel *electronic* computers in quantum chemistry, however, was not explored until several decades later. Work by Clementi and coworkers[3] in the early 1980s demonstrated the use of a loosely coupled array of processors (with 10 compute processors) for computation of Hartree-Fock energies for a short DNA fragment. The Hartree-Fock computations for this 87-atom system, using a basis set with 315 basis functions, represented an extraordinary feat at the time. Other early work exploring the use of parallel computers in quantum chemistry include parallel implementations of both the Hartree-Fock method and a few correlated electronic structure methods. Thus, by the late 1980s, parallel programs had been developed for computation of Hartree-Fock energies[4,5] and gradients,[6] transformation of the two-electron integrals,[4,5,7] and computation of energies with second-order Møller-Plesset perturbation theory[5,8] (MP2) and the configuration interaction (CI) method.[4]

To a large extent, these pioneering efforts employed custom-built parallel computers such as the loosely coupled arrays of processors (LCAP),[3,4,6,8] although one of the early commercially available parallel computers, the Intel hypercube, was used as well.[5,7] In the late 1980s, the increasing availability of commercial parallel computers and the concomitant improvements in the requisite software, such as message-passing libraries, spurred many research groups to undertake development of parallel quantum chemistry applications. The development of parallel quantum chemistry methods, including the earliest efforts as well as later works carried out until the mid-1990s, has been reviewed elsewhere.[9,10] By the late 1990s, great strides had been made toward extending parallel computing in quantum chemistry to a wider range of methods and procedures, for instance, Hartree-Fock second derivatives,[11] MP2 gradients,[12] multireference CI,[13] multireference pseudospectral CI,[14] full CI,[15] and the coupled-cluster method.[16,17]

At present, parallel computing in quantum chemistry continues to be an active field of research; new and improved parallel algorithms for well-established quantum chemical methods are steadily appearing in the literature, and reports of new computational methods are often followed by development of algorithms for their parallel execution. Examples of improved parallel algorithms for previously parallelized methods include a recent parallel implementation of the full CI method,[18] which enabled full CI calculations of an unprecedented size and provided valuable benchmark data, and a novel parallel implementation of the resolution-of-the-identity MP2 method, demonstrating computation of energies and gradients for large systems;[19] also, the first parallel implementation of the MP2-R12 method has been reported.[20] Furthermore, parallel algorithms have been developed for linear-scaling methods, for instance, linear-scaling computation of the Fock matrix,[21,22] which is the major computational step in Hartree-Fock and density functional theory, and the local MP2 method has been parallelized as well.[23] Additionally, new directions are being explored, such as automatic parallel code generation by the Tensor Contraction Engine[24,25] and utilization of multilevel parallelism to achieve improved parallel efficiency.[26]

These research efforts have resulted in a variety of quantum chemistry program packages with parallel capabilities; indeed, most quantum chemistry software suites today support some level of parallelism. A number of the parallel quantum chemistry program suites are freely available, and there are several commercial programs as well. Among the free program packages are: COLUMBUS,[27] which has parallelized versions of the multireference CI and multireference average quadratic coupled-cluster methods; DALTON,[28] which includes parallel Hartree-Fock and density functional theory (DFT); GAMESS (US),[29] with parallelized Hartree-Fock, DFT, and correlated wave functions, up to second derivatives for certain methods; MPQC,[30] designed to be parallel from the beginning, with parallel Hartree-Fock, DFT, and MP2 energies and gradients and parallel explicitly correlated MP2 energies; and NWChem,[31] also incorporating parallelism from the onset, supporting a wide range of methods including Hartree-Fock, DFT, time-dependent DFT, coupled-cluster methods, CI methods, as well as high-level methods for properties. Several commercial quantum chemistry program packages incorporate varying levels of parallelism; among these are ACES III,[32] GAMESS (UK),[33] Gaussian,[34] MOLCAS,[35] MOLPRO,[36] and Q-Chem.[37]

1.2 Trends in Hardware Development

Let us briefly look at some trends in hardware technology that have consequences for developers of parallel scientific applications and for the efficient utilization of single processors as well.

1.2.1 Moore's Law

A central trend in hardware development is the exponential increase with time of the number of transistors in an integrated circuit. This observation, known as Moore's law, was first made in 1965 by Gordon Moore,[38] who found that the number of transistors that minimized the cost per component in an

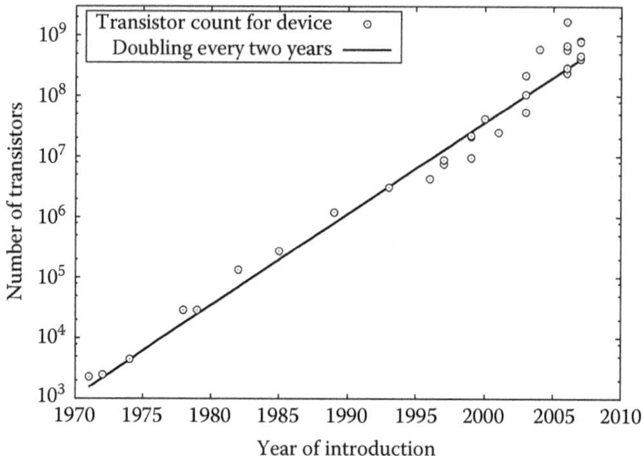

FIGURE 1.1
Historical development of the transistor count for integrated circuits. The solid line depicts Moore's 1975 prediction that transistor counts would double every two years. Data were obtained from Intel's web site, http://www.intel.com, and from Wikipedia, http://www.wikipedia.org.

integrated circuit was doubling every year and predicted that this trend would continue until 1975. In 1975, Moore updated his prediction,[39] estimating that the number of transistors in an integrated circuit would double every two years; three decades later, this trend still continues as illustrated in Figure 1.1.

1.2.2 Clock Speed and Performance

The performance of an integrated circuit depends on the number of transistors, which has increased following Moore's law, and also on the *clock speed*, that is, the rate at which the circuit performs its most fundamental operations. Figure 1.2 shows the development over time of the integer and floating point performance of single chips along with the clock speeds of those chips. It is apparent from the figure that the performance improvement does not parallel that of the clock speed. The development of the clock speed relative to the instruction rate falls into three periods of chip development. The first period is distinguished by the use of *pipelining*, enabling instructions to be performed in several stages that are overlapped to improve performance. This is marked by the rapid improvement in performance relative to clock speed up until around 1989. During the second period, *multiple-issue* (or *superscalar*) processors were developed and improved. Such processors allow more than one instruction to start execution on each clock cycle. During this period the achieved performance gains were affected by a variety of factors, and the lack of equivalent gains in bandwidth and latency (discussed in the next section) resulted in an overall decrease in performance achieved per clock cycle. The third, and current, period is that of *multicore* processors. Clock speeds in these processors have dropped from their peak, yet significant performance

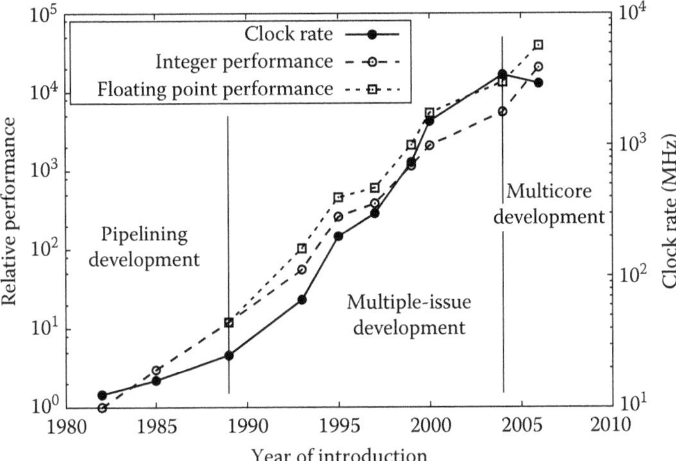

FIGURE 1.2
Development of clock rates, integer performance, and floating point performance for selected Intel chips from 1982 until the present. Integer performance data were normalized to a value of 1 for the first data point (1982); floating point performance data were normalized to make the first data point (1989) match the integer performance for that year. Integer performance was derived from SPECint92, SPECint95, SPECint2000, and SPECint2000rate benchmarks,[50] using the latest benchmark available for a given chip. For the oldest two chips, the maximum theoretical rate for instruction execution was employed, and the availability of several benchmarks for some processors allowed normalization of the data so that performance data from different benchmarks could be used. A similar approach was used for floating point performance using the floating point versions of the SPEC benchmarks. Data were obtained from Wikipedia, http://www.wikipedia.org; Intel's web site, http://www.intel.com; the SPEC web site, http://www.spec.org; and Hennessy and Patterson.[51] Various sources found with the Google news archive at http://news.google.com assisted in determining the year of introduction of each chip.

improvements are still realized by essentially completely replicating multiple processors on the same chip. The trend of placing multiple processors on a single integrated circuit is expected to continue for some time into the future.

1.2.3 Bandwidth and Latency

Important factors determining processor performance include the memory bandwidth and latency. The memory bandwidth is the rate at which a processor can transfer data to or from memory, and the memory latency is the time that elapses between a request for a data transfer and the arrival of the first datum. Figure 1.3 illustrates the development of the memory bandwidth and latency relative to the floating point performance over the last two decades. For any given year, the figure shows the ratio of the memory bandwidth and the inverse latency to the floating point performance, and both ratios have been normalized to yield a value of 1 in 1989; note that no absolute performance data can be gleaned from the figure. From the downward trends of the curves, however, it is clear that the improvements in latency lag the

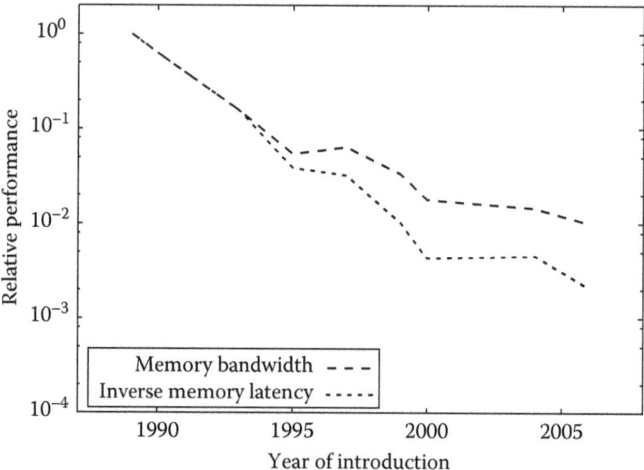

FIGURE 1.3
Development of memory bandwidth and inverse latency relative to processor floating point performance from 1989 to the present. For a given year, the figure shows the memory bandwidth and inverse latency divided by the floating point performance, and the data are normalized to yield a value of 1 for 1989. Floating point performance data are identical to those reported in Figure 1.2; memory bandwidth and latency data were taken from Ref. 40 and, for recent years, obtained from Wikipedia, http://www.wikipedia.org, and with the assistance of the Google news archive, http://news.google.com.

improvements in bandwidth, which, in turn, are not as fast as the increase in the floating point performance. Similar trends are observed in bandwidths and latencies for networks and disks as well.[40]

To improve the bandwidth and reduce the latency, most modern processors can execute instructions out of order, permitting instructions to be scheduled ahead of those that are awaiting data. Also, a substantial portion of the available transistors on a chip are now dedicated to caches that hold frequently used data closer to the processor to reduce the latency penalty for each access. These measures, however, have not been sufficient to completely offset the effects of the relative decrease in the speeds of data movement. The much faster improvement in peak floating point computation rates compared to improvements in both memory and network bandwidth and latency have made efficient use of modern computers, in particular parallel computers, more challenging. For example, a parallel algorithm that had a negligible communication overhead when developed a decade ago may have a reduced parallel performance on today's computers because the faster improvement in floating point performance has made the communication time of the algorithm nonnegligible compared to the computation time.

1.2.4 Supercomputer Performance

In addition to steady gains in the performance of a single chip, the speeds of the fastest parallel computers have shown tremendous growth. The TOP500

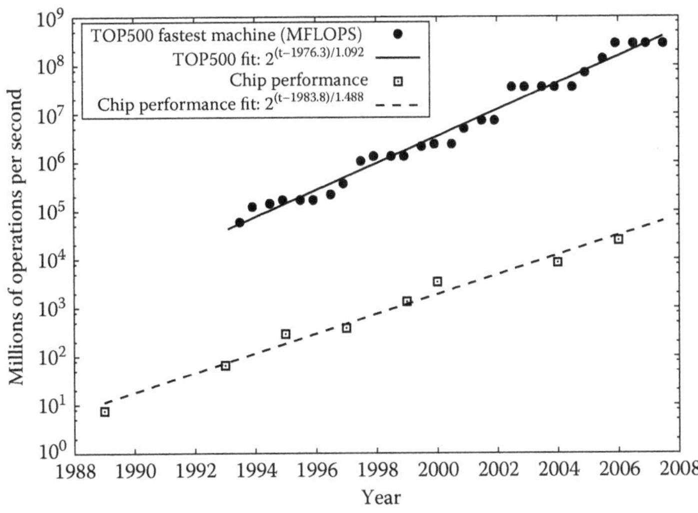

FIGURE 1.4
Semilogarithmic plot of the development over time of the speed of the world's fastest supercomputer and the floating point performance of a single chip. The speed of the fastest parallel computer doubles about every year, while the processing speed of a chip doubles about every eighteen months. The fastest supercomputer speed is measured as the floating point rate (in millions of floating point operations per second, MFLOPS) for the HPL benchmark for the computer at the top of the TOP500 list.[41] The single-chip floating point performance is normalized to the effective single-chip floating point performance in the HPL benchmark determined for the Sandia Thunderbird computer (described in section 2.4.5) for which both HPL and single-chip data were available.

list,[41] compiled every year since 1993, ranks the world's fastest parallel computers using the High-Performance Linpack benchmark[42] (HPL). Figure 1.4 shows the development over time of the speed of the fastest TOP500 computer and the floating point performance of the fastest single chip. The speed of the fastest parallel computer, doubling roughly every year, grows at a significantly faster rate than the speed of the fastest chip, which doubles about every eighteen months. This difference in growth rates is due mostly to the increase over the years in the total number of chips contained by the largest parallel computers. If these trends hold, the fastest TOP500 supercomputer will run at a speed of around 7 petaflops in the year 2012.

Regarding single-chip performance, given the stagnation in clock speeds shown in Figure 1.2, it is reasonable to expect that most of the on-chip improvements in the near future will be obtained by adding more processor cores. The 2006 chip performance data in Figure 1.4 were obtained for a two-core chip, and, extrapolating the trend in floating point performance improvements, we would then expect each chip to have around 32 processor cores in 2012. Alternatively, one might find that chip manufacturers would switch to more, but simpler, cores on a chip, and this could dramatically increase the number of cores available. As a case in point, a research chip containing 80 cores was demonstrated by Intel in 2006.

1.3 Trends in Parallel Software Development

In the previous section we briefly discussed some of the major current trends in hardware development, and below we will consider the effects of this development on software design. Moreover, we will touch upon how the evolution of methods and algorithms in itself affects the parallelization of applications. The combined effect of the advances in hardware and software is a rapidly evolving computational environment that poses a serious challenge for the program developer wishing to design long-lasting parallel software that will continue to take advantage of hardware performance gains; we will conclude this section with a discussion of new programming models that are designed to address this challenge.

1.3.1 Responding to Changes in Hardware

In the past, efficient utilization of large-scale parallel computers has required adaptation of applications to accommodate a number of processors that doubled every three years. If the current trends continue to hold, however, the number of processors on parallel computers is expected to grow even faster, doubling every year. Thus, writing scalable applications, namely, parallel programs that can take advantage of a very large number of processors, becomes increasingly important. For example, this may entail putting more emphasis on reducing the communication overhead when designing parallel applications; using a significant amount of collective communication in a parallel application usually precludes scalability, and, moreover, the slower improvements in network performance than in instruction rates make it increasingly difficult to hide the communication overhead.

Although the rapidly increasing number of processors on the largest supercomputers emphasizes the need for efficient massively parallel applications, the increasing availability of small-scale parallel computers makes small-scale parallelism important as well. Small-scale parallelism is usually considerably easier to incorporate into a scientific application; achieving high parallel efficiency on a small number, perhaps a few tens, of processors may require relatively modest modifications to existing scalar code, whereas high parallel efficiency on thousands of processors requires a very carefully designed parallel application.

1.3.2 New Algorithms and Methods

Considering quantum chemistry in particular, hardware improvements have made it possible to tackle different classes of problems, posing a new set of challenges for parallel program developers. For instance, parallelization of conventional quantum chemistry methods, whose computational cost scales as a high-degree polynomial in the molecular size, is rather forgiving in terms of what steps are parallelized. Consider, for example, conventional MP2 theory, whose computational cost is dominated by $O(N^5)$ and $O(N^4)$ steps for large N, where N is the size of the molecule. If the $O(N^5)$ steps can be perfectly parallelized without communication overhead and the $O(N^4)$ steps are run sequentially, then the time to solution, as a function of N and the number of processors, p, can be expressed as

$$t_{\mathrm{MP2}}(N, p) \approx A\frac{N^5}{p} + BN^4 \tag{1.1}$$

where A and B are constants and one process is run per processor. The fraction of time spent in serial code when running with a single process is $f_{\mathrm{MP2}} \approx B/(AN+B)$, and from Amdahl's law (discussed in section 5.2.1) it then follows that the maximum attainable speedup is

$$S_{\mathrm{max,MP2}} = \frac{1}{f_{\mathrm{MP2}}} \approx \frac{AN+B}{B}. \tag{1.2}$$

The parallel efficiency (also defined in section 5.2.1), which measures how well the computer is utilized, is then given as

$$E_{\mathrm{MP2}}(N, p) = \frac{t_{\mathrm{MP2}}(N, 1)}{p t_{\mathrm{MP2}}(N, p)} \approx \frac{1}{1 + \frac{(p-1)}{AN/B+1}} \tag{1.3}$$

and a perfect parallelization corresponds to $E_{\mathrm{MP2}}(N, p) = 1$. From Eq. 1.2 it follows that the maximum attainable speedup increases with the problem size. Also, we see from Eq. 1.3 that the efficiency decreases with the number of processors, but if the problem size increases at the same rate as the number of processors, the efficiency remains nearly constant. Thus, we can utilize a large number of processors with reasonably high efficiency provided that we have a large enough problem to solve. However, for a linear-scaling method, such as LMP2, we do not have such a luxury. Assuming that we have a linear-scaling algorithm with two linear-scaling terms, only one of which is parallelized, the time to solution is

$$t_{\mathrm{LMP2}}(N, p) \approx C\frac{N}{p} + DN \tag{1.4}$$

where C and D are constants. This yields the following maximum speedup and efficiency:

$$S_{\mathrm{max,LMP2}} \approx \frac{C+D}{D} \tag{1.5}$$

$$E_{\mathrm{LMP2}}(N, p) \approx \frac{1}{1 + \frac{(p-1)}{C/D+1}}. \tag{1.6}$$

In this case, there is a fixed upper limit to the achievable speedup, and the efficiency decreases as the number of processors increases, regardless of the problem size. Thus, much work must be invested to achieve efficient parallelization of linear-scaling methods; these methods involve a large number of steps that scale linearly in N, and each step must be parallelized to achieve high efficiency when running on large numbers of processors.

1.3.3 New Programming Models

In this book we will focus on the traditional message-passing programming model typically employed in parallel applications, as well as hybrid programming models, where message-passing is combined with multi-threading. However, given the continual rapid advances in hardware, algorithms, and methods, it is worthwhile to reconsider whether the traditional and hybrid programming models are the best approaches for obtaining high performance. Our considerations should encompass the entire software life cycle, including the cost of writing, debugging, porting, and extending the application program. It is likely that alternative programming models will be able to provide ways to express parallelism that are simpler and less error-prone and yield higher efficiency on large parallel computers than the traditional models.

The partitioned global address space (PGAS) languages provide an alternative to traditional message-passing programming models that allows programmers to view the distributed memory of a parallel machine as if the memory were shared and directly addressable from all of the processes executing the application, subject to various constraints. The application can take advantage of locality to achieve high performance because mechanisms are provided for the process to determine which memory is local and which is remote (local memory access is faster than remote memory access). Several PGAS languages exist, and examples include: Unified Parallel C[43] (UPC), which is derived from C; Co-array Fortran,[44] derived from Fortran; and Titanium,[45] derived from the Java™ language. The hybrid programming examples presented later in this book as well as the Global Arrays package[46] use the PGAS concept, but employ software libraries to implement data sharing rather than having the explicit language support that is provided by the PGAS languages.

Programming model research is also being conducted by the Defense Advanced Research Projects Agency (DARPA) under the High Productivity Computing Systems (HPCS) program. The desired outcome of this program is the creation of computing environments that run application programs more efficiently than existing architectures while keeping the time and effort required to write applications for the new architectures to a minimum. A component of this research is the development of languages that contain advanced parallel computing constructs. The languages that have been created by the DARPA HPCS program, and are still being researched, are X10,[47] Chapel,[48] and Fortress.[49] As of the time of writing, these efforts hold promise but are not ready for adoption.

References

1. Löwdin, P. O. Some aspects on the history of computational quantum chemistry in view of the development of the supercomputers and large-scale parallel computers. In M. Dupuis (Ed.), *Lecture Notes in Chemistry*, No. 44, *Supercomputer Simulations in Chemistry*, pp. 1–48. Berlin-Heidelberg: Springer-Verlag, 1986.
2. Fröman, A., and J. Linderberg. *Inception of Quantum Chemistry at Uppsala*. Uppsala, Sweden: Uppsala University Press, 2007.
3. Clementi, E., G. Corongiu, J. Detrich, S. Chin, and L. Domingo. Parallelism in quantum chemistry: Hydrogen bond study in DNA base pairs as an example. *Int. J. Quant. Chem. Symp.* 18:601–618, 1984.
4. Guest, M. F., R. J. Harrison, J. H. van Lenthe, and L. C. H. van Corler. Computational chemistry on the FPS-X64 scientific computers. *Theor. Chim. Acta* 71:117–148, 1987.
5. Colvin, M. E., R. A. Whiteside, and H. F. Schaefer III. In S. Wilson (Ed.), *Methods in Computational Chemistry* III: 167–237, 1989.
6. Dupuis, M., and J. D. Watts. Parallel computation of molecular energy gradients on the loosely coupled array of processors (LCAP). *Theor. Chim. Acta* 71:91–103, 1987.
7. Whiteside, R. A., J. S. Binkley, M. E. Colvin, and H. F. Schaefer III. Parallel algorithms for quantum chemistry. I. Integral transformations on a hypercube multiprocessor. *J. Chem. Phys.* 86:2185–2193, 1987.
8. Watts, J. D., and M. Dupuis. Parallel computation of the Møller–Plesset second-order contribution to the electronic correlation energy. *J. Comp. Chem.* 9:158–170, 1988.
9. Harrison, R. J., and R. Shepard. Ab initio molecular electronic structure on parallel computers. *Ann. Rev. Phys. Chem.* 45:623–658, 1994.
10. Kendall, R. A., R. J. Harrison, R. J. Littlefield, and M. F. Guest. High performance computing in computational chemistry: Methods and machines. *Rev. Comp. Chem.* 6:209–316, 1995.
11. Márquez, A. M., J. Oviedo, J. F. Sanz, and M. Dupuis. Parallel computation of second derivatives of RHF energy on distributed memory computers. *J. Comp. Chem.* 18:159–168, 1997.
12. Nielsen, I. M. B. A new direct MP2 gradient algorithm with implementation on a massively parallel computer. *Chem. Phys. Lett.* 255:210–216, 1996.
13. Dachsel, H., H. Lischka, R. Shepard, J. Nieplocha, and R. J. Harrison. A massively parallel multireference configuration interaction program: The parallel COLUMBUS program. *J. Comp. Chem.* 18:430–448, 1997.
14. Martinez, T., and E. A. Carter. Pseudospectral correlation methods on distributed memory parallel architectures. *Chem. Phys. Lett.* 241:490–496, 1995.
15. Evangelisti, S., G. L. Bendazzoli, R. Ansaloni, and E. Rossi. A full CI algorithm on the CRAY T3D. Application to the NH_3 molecule. *Chem. Phys. Lett.* 233:353–358, 1995.
16. Rendell, A. P., T. J. Lee, and R. Lindh. Quantum chemistry on parallel computer architectures: Coupled-cluster theory applied to the bending potential of fulminic acid. *Chem. Phys. Lett.* 194:84–94, 1992.
17. Kobayashi, R., and A. P. Rendell. A direct coupled cluster algorithm for massively parallel computers. *Chem. Phys. Lett.* 265:1–11, 1997.

18. Gan, Z., and R. J. Harrison. Calibrating quantum chemistry: A multiteraflop, parallel-vector, full-configuration interaction program for the Cray-X1. In *SC '05: Proceedings of the 2005 ACM/IEEE Conference on Supercomputing*, p. 22. Washington, DC: IEEE Computer Society, 2005.
19. Hättig, C., A. Hellweg, and A. Köhn. Distributed memory parallel implementation of energies and gradients for second-order Møller–Plesset perturbation theory with the resolution-of-the-identity approximation. *Phys. Chem. Chem. Phys.* 8:1159–1169, 2006.
20. Valeev, E. F., and C. L. Janssen. Second-order Møller–Plesset theory with linear R12 terms (MP2-R12) revisited: Auxiliary basis set method and massively parallel implementation. *J. Chem. Phys.* 121:1214–1227, 2004.
21. Gan, C. K., C. J. Tymczak, and M. Challacombe. Linear scaling computation of the Fock matrix. VII. Parallel computation of the Coulomb matrix. *J. Chem. Phys.* 121:6608–6614, 2004.
22. Weber, V., and M. Challacombe. Parallel algorithm for the computation of the Hartree-Fock exchange matrix: Gas phase and periodic parallel ONX. *J. Chem. Phys.* 125:104110, 2006.
23. Nielsen, I. M. B., and C. L. Janssen. Local Møller–Plesset perturbation theory: A massively parallel algorithm. *J. Chem. Theor. Comp.* 3:71–79, 2007.
24. Hirata, S. Tensor Contraction Engine: Abstraction and automated parallel implementation of configuration-interaction, coupled-cluster, and many-body perturbation theories. *J. Phys. Chem. A* 107:9887–9897, 2003.
25. Auer, A. A., G. Baumgartner, D. E. Bernholdt, A. Bibireata, V. Choppella, D. Cociorva, X. Gao, R. Harrison, S. Krishnamoorthy, S. Krishnan, C.-C. Lam, Q. Lu, M. Nooijen, R. Pitzer, J. Ramanujam, P. Sadayappan, and A. Sibiryakov. Automatic code generation for many-body electronic structure methods: The Tensor Contraction Engine. *Mol. Phys.* 104:211–228, 2006.
26. Krishnan, M., Y. Alexeev, T. L. Windus, and J. Nieplocha. Multilevel parallelism in computational chemistry using Common Component Architecture and Global Arrays. In *SC '05: Proceedings of the 2005 ACM/IEEE Conference on Supercomputing*, p. 23. Washington, DC: IEEE Computer Society, 2005.
27. Lischka, H., R. Shepard, I. Shavitt, R. M. Pitzer, M. Dallos, T. Müller, P. G. Szalay, F. B. Brown, R. Ahlrichs, H. J. Böhm, A. Chang, D. C. Comeau, R. Gdanitz, H. Dachsel, C. Ehrhardt, M. Ernzerhof, P. Höchtl, S. Irle, G. Kedziora, T. Kovar, V. Parasuk, M. J. M. Pepper, P. Scharf, H. Schiffer, M. Schindler, M. Schüler, M. Seth, E. A. Stahlberg, J.-G. Zhao, S. Yabushita, Z. Zhang, M. Barbatti, S. Matsika, M. Schuurmann, D. R. Yarkony, S. R. Brozell, E. V. Beck, and J.-P. Blaudeau. COLUMBUS, an ab initio electronic structure program, release 5.9.1 (2006). http://www.univie.ac.at/columbus/.
28. DALTON, a molecular electronic structure program, Release 2.0 (2005). http://www.kjemi.uio.no/software/dalton/dalton.html.
29. Schmidt, M. W., K. K. Baldridge, J. A. Boatz, S. T. Elbert, M. S. Gordon, J. H. Jensen, S. Koseki, N. Matsunaga, K. A. Nguyen, S. J. Su, T. L. Windus, M. Dupuis, and J. A. Montgomery. General Atomic and Molecular Electronic Structure System. *J. Comp. Chem.* 14:1347–1363, 1993. http://www.msg.chem.iastate.edu/gamess/.
30. Janssen, C. L., I. B. Nielsen, M. L. Leininger, E. F. Valeev, J. Kenny, and E. T. Seidl. The Massively Parallel Quantum Chemistry program (MPQC), version 3.0.0-alpha, Sandia National Laboratories. Livermore, CA, 2007. http://www.mpqc.org.

31. Bylaska, E. J., W. A. de Jong, K. Kowalski, T. P. Straatsma, M. Valiev, D. Wang, E. Aprà, T. L. Windus, S. Hirata, M. T. Hackler, Y. Zhao, P.-D. Fan, R. J. Harrison, M. Dupuis, D. M. A. Smith, J. Nieplocha, V. Tipparaju, M. Krishnan, A. A. Auer, M. Nooijen, E. Brown, G. Cisneros, G. I. Fann, H. Früchtl, J. Garza, K. Hirao, R. Kendall, J. A. Nichols, K. Tsemekhman, K. Wolinski, J. Anchell, D. Bernholdt, P. Borowski, T. Clark, D. Clerc, H. Dachsel, M. Deegan, K. Dyall, D. Elwood, E. Glendening, M. Gutowski, A. Hess, J. Jaffe, B. Johnson, J. Ju, R. Kobayashi, R. Kutteh, Z. Lin, R. Littlefield, X. Long, B. Meng, T. Nakajima, S. Niu, L. Pollack, M. Rosing, G. Sandrone, M. Stave, H. Taylor, G. Thomas, J. van Lenthe, A. Wong, and Z. Zhang. NWChem, a computational chemistry package for parallel computers, version 5.0 (2006), Pacific Northwest National Laboratory, Richland, Washington 99352-0999, USA. http://www.emsl.pnl.gov/docs/nwchem/nwchem.html.
32. Lotrich, V., M. Ponton, L. Wang, A. Yau, N. Flocke, A. Perera, E. Deumens, and R. Bartlett. The super instruction processor parallel design pattern for data and floating point intensive algorithms. Workshop on Patterns in High Performance Computing, May 2005. ACES III is a parallelized version of ACES II. ACES II is a program product of the Quantum Theory Project, University of Florida. Authors: J. F. Stanton, J. Gauss, J. D. Watts, M. Nooijen, N. Oliphant, S. A. Perera, P. G. Szalay, W. J. Lauderdale, S. A. Kucharski, S. R. Gwaltney, S. Beck, A. Balková D. E. Bernholdt, K. K. Baeck, P. Rozyczko, H. Sekino, C. Hober, and R. J. Bartlett. Integral packages included are VMOL (J. Almlöf and P. R. Taylor); VPROPS (P. Taylor); ABACUS (T. Helgaker, H. J. Aa. Jensen, P. Jørgensen, J. Olsen, and P. R. Taylor).
33. Guest, M. F., I. J. Bush, H. J. J. van Dam, P. Sherwood, J. M. H. Thomas, J. H. van Lenthe, R. W. A Havenith, J. Kendrick. The GAMESS-UK electronic structure package: algorithms, developments and applications. *Mol. Phys.* 103:719–747, 2005. http://www.cfs.dl.ac.uk/gamess-uk/index.shtml.
34. Frisch, M. J., G. W. Trucks, H. B. Schlegel, G. E. Scuseria, M. A. Robb, J. R. Cheeseman, J. A. Montgomery, Jr., T. Vreven, K. N. Kudin, J. C. Burant, J. M. Millam, S. S. Iyengar, J. Tomasi, V. Barone, B. Mennucci, M. Cossi, G. Scalmani, N. Rega, G. A. Petersson, H. Nakatsuji, M. Hada, M. Ehara, K. Toyota, R. Fukuda, J. Hasegawa, M. Ishida, T. Nakajima, Y. Honda, O. Kitao, H. Nakai, M. Klene, X. Li, J. E. Knox, H. P. Hratchian, J. B. Cross, V. Bakken, C. Adamo, J. Jaramillo, R. Gomperts, R. E. Stratmann, O. Yazyev, A. J. Austin, R. Cammi, C. Pomelli, J. W. Ochterski, P. Y. Ayala, K. Morokuma, G. A. Voth, P. Salvador, J. J. Dannenberg, V. G. Zakrzewski, S. Dapprich, A. D. Daniels, M. C. Strain, O. Farkas, D. K. Malick, A. D. Rabuck, K. Raghavachari, J. B. Foresman, J. V. Ortiz, Q. Cui, A. G. Baboul, S. Clifford, J. Cioslowski, B. B. Stefanov, G. Liu, A. Liashenko, P. Piskorz, I. Komaromi, R. L. Martin, D. J. Fox, T. Keith, M. A. Al-Laham, C. Y. Peng, A. Nanayakkara, M. Challacombe, P. M. W. Gill, B. Johnson, W. Chen, M. W. Wong, C. Gonzalez, and J. A. Pople. Gaussian 03. Gaussian, Inc., Wallingford, CT, 2004.
35. Karlström, G., R. Lindh, P.-A. Malmqvist, B. O. Roos, U. Ryde, V. Veryazov, P.-O. Widmark, M. Cossi, B. Schimmelpfening, P. Neogrady, and L. Seijo. Molcas: a program package for computational chemistry. *Comp. Mat. Sci.* 28:222–239, 2003.
36. MOLPRO, version 2006.1, a package of ab initio programs, H.-J. Werner, P. J. Knowles, R. Lindh, F. R. Manby, M. Schütz, P. Celani, T. Korona, G. Rauhut, R. D. Amos, A. Bernhardsson, A. Berning, D. L. Cooper, M. J. O. Deegan, A. J. Dobbyn, F. Eckert, C. Hampel and G. Hetzer, A. W. Lloyd, S. J. McNicholas, W. Meyer and M. E. Mura, A. Nicklass, P. Palmieri, R. Pitzer, U. Schumann, H. Stoll, A. J. Stone, R. Tarroni and T. Thorsteinsson. http://www.molpro.net.

37. Shao, Y., L. Fusti-Molnar, Y. Jung, J. Kussmann, C. Ochsenfeld, S. T. Brown, A. T. B. Gilbert, L. V. Slipchenko, S. V. Levchenko, D. P. O'Neill, R. A. D. Jr., R. C. Lochan, T. Wang, G. J. O. Beran, N. A. Besley, J. M., Herbert, C. Y. Lin, T. Van Voorhis, S. H. Chien, A. Sodt, R. P. Steele, V. A. Rassolov, P. E. Maslen, P. P. Korambath, R. D. Adamson, B. Austin, J. Baker, E. F. C. Byrd, H. Dachsel, R. J. Doerksen, A. Dreuw, B. D. Dunietz, A. D. Dutoi, T. R. Furlani, S. R. Gwaltney, A. Heyden, S. Hirata, C.-P. Hsu, G. Kedziora, R. Z. Khalliulin, P. Klunzinger, A. M. Lee, M. S. Lee, W. Liang, I. Lotan, N. Nair, B. Peters, E. I. Proynov, P. A. Pieniazek, Y. M. Rhee, J. Ritchie, E. Rosta, C. D. Sherrill, A. C. Simmonett, J. E. Subotnik, H. L. Woodcock III, W. Zhang, A. T. Bell, A. K. Chakraborty, D. M. Chipman, F. J. Keil, A. Warshel, W. J. Hehre, H. F. Schaefer III, J. Kong, A. I. Krylov, P. M. W. Gill, and M. Head-Gordon. Advances in methods and algorithms in a modern quantum chemistry program package. *Phys. Chem. Chem. Phys.* 8:3172–3191, 2006.
38. Moore, G. E. Cramming more components onto integrated circuits. *Electronics* 38, 1965.
39. Moore, G. E. Progress in digital integrated electronics. In *Technical Digest 1975. International Electron Devices Meeting*, pp. 11–13. IEEE Computer Society, 1975.
40. Patterson, D. A. Latency lags bandwidth. *Commun. ACM* 47, 2004.
41. Meuer, H., E. Strohmaier, J. Dongarra, and H. D. Simon. TOP500 Supercomputing Sites. June 1993 to June 2007. http://www.top500.org.
42. Petitet, A., R. C. Whaley, J. Dongarra, and A. Cleary. HPL—a portable implementation of the high-performance Linpack benchmark for distributed-memory computers. http://www.netlib.org/benchmark/hpl/.
43. El-Ghazawi, T., W. Carlson, T. Sterling, and K. Yelick. *UPC: Distributed Shared-Memory Programming*. Hoboken, NJ: John Wiley & Sons, 2005.
44. Numrich, R. W., and J. Reid. Co-arrays in the next Fortran standard. *SIGPLAN Fortran Forum* 24(2):4–17, 2005.
45. Yelick, K., P. Hilfinger, S. Graham, D. Bonachea, J. Su, A. Kamil, K. Datta, P. Colella, and T. Wen. Parallel languages and compilers: Perspective from the Titanium experience. *Int. J. High Perform. C.* 21:266–290, 2007.
46. Nieplocha, J., R. J. Harrison, and R. J. Littlefield. Global arrays: A portable "shared-memory" programming model for distributed memory computers. In *Proceedings of the 1994 conference on Supercomputing*, pp. 340–349. Los Alamitos: IEEE Computer Society Press, 1994.
47. Charles, P., C. Grothoff, V. Saraswat, C. Donawa, A. Kielstra, K. Ebcioglu, C. von Praun, and V. Sarkar. X10: an object-oriented approach to nonuniform cluster computing. In OOPSLA '05: *Proceedings of the 20th Annual ACM SIGPLAN Conference on Object Oriented Programming, Systems, Languages, and Applications*, pp. 519–538. New York: ACM Press, 2005.
48. Chamberlain, B. L., D. Callahan, and H. P. Zima. Parallel programmability and the Chapel language. *Int. J. High Perform. C.* 21:291–312, 2007.
49. Allen, E., D. Chase, J. Hallett, V. Luchangco, J.-W. Maessen, S. Ryu, J. G. L. Steele, and S. Tobin-Hochstadt. *The Fortress Language Specification: Version 1.0 β*. San Francisco: Sun Microsystems, 2007.
50. The Standard Performance Evaluation Corporation, SPEC, has developed benchmark suites to quantity the performance of computers executing tasks in a variety of application areas. More information and results can be found at http://www.spec.org.
51. Hennessy J. L., and D. A. Patterson. *Computer Architecture: A Quantitative Approach*, 4th edition. San Francisco: Morgan Kaufmann, 2007.

2

Parallel Computer Architectures

A basic understanding of parallel processing hardware is required to design and implement parallel software that gets the best performance out of a computer and to choose a suitable hardware platform for running a particular application. In this chapter, we will first introduce various computer architectures, using Flynn's taxonomy. We will then discuss the two main components defining a parallel computer, namely, the nodes and the network connecting the nodes, where a node is a set of processors grouped together and sharing memory and other resources. The discussion of parallel network architecture will compare the properties of various topologies for the networks connecting the nodes. Finally, we will provide a detailed discussion of the multiple-instruction, multiple-data (MIMD) architecture, which has emerged as the dominant architecture for large-scale computers, covering MIMD memory organization, reliability, and impact on parallel program design.

2.1 Flynn's Classification Scheme

Parallel computers are often classified according to a scheme proposed by Michael Flynn in 1972.[1] In the Flynn taxonomy, illustrated in Table 2.1, there are four classes of parallel computers, distinguished on the basis of the flow of data and instructions: an application running on a computer is viewed as one or more sequences (or "streams") of instructions and one or more streams of data, and computers are divided into four classes depending on whether multiple streams of data or instructions are permitted. Three of these four classes of parallel computer architectures are of interest for quantum chemistry applications and are discussed in detail next.

2.1.1 Single-Instruction, Single-Data

The *single-instruction, single-data* (SISD) architecture has a single instruction stream and a single data stream. A single processor in a personal computer serves as an example of this type of architecture. The instruction stream and

TABLE 2.1

Flynn's taxonomy for parallel computers

SISD	Single-instruction, single-data
SIMD	Single-instruction, multiple-data
MISD	Multiple-instruction, single-data
MIMD	Multiple-instruction, multiple-data

the data stream interact, and the data stream is entirely determined by those instructions in the instruction stream that pertain to the movement of data. We will discuss SISD processors in more detail in section 2.3, where we consider their role as components in a parallel computer.

2.1.2 Single-Instruction, Multiple-Data

Scientific computing often involves performing the same operation on different pieces of data. A natural way to do this is provided by the *single-instruction, multiple-data* (SIMD) architecture, in which a single instruction stream simultaneously operates on multiple data streams. The most extreme example of a SIMD architecture is the Connection Machine,[2] the first models of which appeared in the 1980s. In the CM-2 model of this computer, a single instruction stream is decoded and used to control the operation of up to 65,536 computing units, each processing its own data stream.

More limited versions of SIMD architectures are common today. Rather than simultaneously computing all results as is done in the CM-2, the more modern SIMD architectures pipeline data through one or more execution units and can achieve high computational rates due to the repetitiveness of the work involved. This arrangement is used in vector processors (for example, the Cray X1™). Also, modern microprocessors now have extensions allowing single instructions to perform identical operations on multiple pieces of data. Examples include Intel's Streaming SIMD Extensions (SSE) and Motorola® AltiVec™ used in recent versions of the IBM POWER™ processor.

Large-scale SIMD computers such as the CM-2 have not survived in the marketplace because a limited range of applications are amenable to the employed programming model and because of the continued trend, as predicted by Moore's law, lowering the cost of making each computational unit a full-featured device with its own instruction decoder.

2.1.3 Multiple-Instruction, Multiple-Data

The *multiple-instruction, multiple-data* (MIMD) architecture permits multiple instruction streams to simultaneously interact with their own data stream. While MIMD machines composed of completely independent pairs of instruction and data streams may be of use for trivially parallel applications, it is generally necessary to use a network to connect the processors together in a way that allows a given processor's data stream to be supplemented by data computed by other processors. A MIMD computer of essentially any size can be built by repeatedly adding inexpensive microprocessors and simple network elements as described in section 2.2.4. This feature, combined with programming flexibility, has made MIMD the principal architecture for large-scale parallel computers. The discussion of parallel programming principles and the parallel algorithms presented throughout the remaining chapters of this book will therefore pertain mostly to the MIMD architecture. After the discussion of network and node architectures in the next two sections, we will examine the MIMD architecture in depth in section 2.4.

2.2 Network Architecture

A massively parallel computer consists of a number of nodes connected via a network. A *node* is a group of processors that share memory and other resources, and each node typically contains several processors, although single-processor nodes are used as well. The network connecting the nodes is one of the key components defining a parallel computer. To a large extent, the network determines the performance of the processors collectively running a single parallel job. Moreover, networks are used hierarchically within a parallel computer to connect nodes, processing elements, or other networking elements. We will refer to the endpoints of each of these networks as either *ports* or nodes depending on the context. In this section we will discuss network architecture, including different types of networks, routing of data through a network, and factors determining network performance. Furthermore, we will give an overview of the various topologies of the networks connecting the nodes in large parallel computers, relating the topological properties of the network to the overall machine performance and cost.

2.2.1 Direct and Indirect Networks

Networks may be characterized as either direct or indirect. In a direct network, the computing elements and the network components are combined, and the number of switching elements equals the number of processing elements. The switching elements in a network are the components through which the nodes are connected. Figure 2.1 depicts four computing elements, each with two network links, that form a direct network in a simple ring topology (ring topologies are discussed in section 2.2.4). In an indirect network, the networking and processing elements are separated, allowing more complex, multistage networks to be built. A common example of an indirect network is the fat tree, discussed in section 2.2.4 and shown in Figure 2.2. Any direct network can also be considered as an indirect network with an equal number of separate processing and switching elements. An indirect representation of a ring network is shown in Figure 2.3.

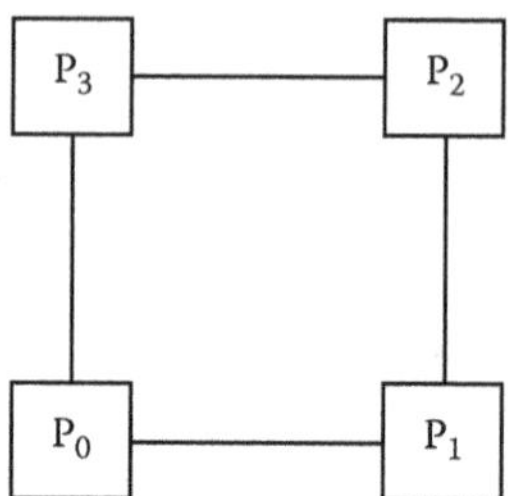

FIGURE 2.1
A direct network with four computing elements, P_i, arranged in a ring.

Networks in modern parallel computers are usually indirect; the computing elements are decoupled from the switching elements, which internally possess the processors and memory required to perform the switching functions. Some modern parallel computers, however, have networks that are direct in the sense that they provide one switching element per node, even though the computing and switching elements are decoupled. For the purpose of performance analysis, we will consider a network with decoupled computing and switching elements to be an indirect network, although those networks that have exactly one switching element per node will be shown in the figures as direct networks with merged computing and switching elements.

2.2.2 Routing

When data is sent between two processing elements, it must typically transverse several switching elements. The networks employed in high-performance computing have enough richness of connectivity that any one of several paths, called *routes*, can be used to move data between two points. Figure 2.4 shows two different routes through a network. A method is needed to select between the possible routes through the network, and a common

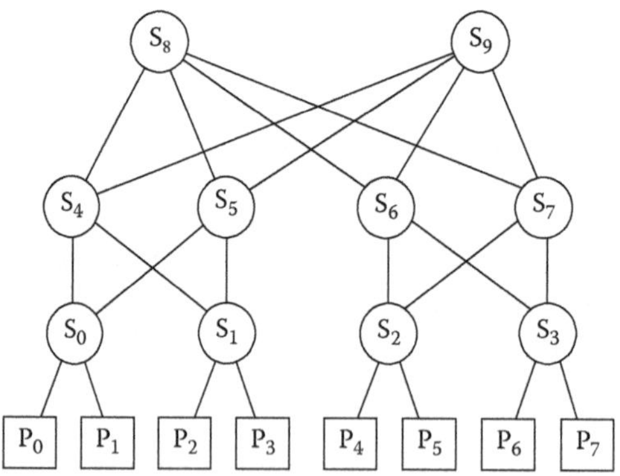

FIGURE 2.2
An indirect network with eight computing elements, P_i, and ten switching elements, S_i. The network shown is a binary three-stage fat tree.

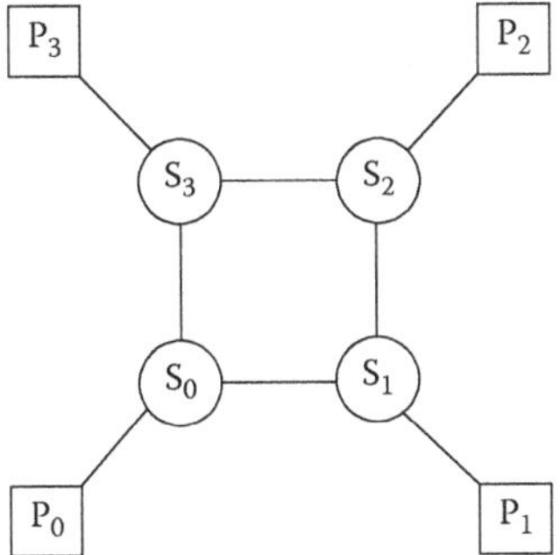

FIGURE 2.3
A ring network shown as an indirect network with separate computing elements, P_i, and network elements, S_i.

approach is *static routing* (also called *oblivious routing*), where the data between two endpoints always follows the same path, regardless of other traffic in the network. Static routing can be accomplished by giving each switch a table that maps the destination address into the next link along which to send the data. Another approach, called *source routing*, entails encoding the sequence of links that the data will follow at the beginning of the data stream.

Each of the routes shown in Figure 2.4 passes through the same number of switches and the same number of links, and one is expected to perform as well as the other when the network is otherwise idle. However, the situation changes when there is more than one pair of nodes involved in data transfers. Figure 2.5 shows network *congestion*, which can occur when additional traffic is present in the network. Here, processing element P_0 is sending data to P_5 using the path $P_0 \rightarrow S_0 \rightarrow S_4 \rightarrow S_9 \rightarrow S_6 \rightarrow S_2 \rightarrow P_5$, and P_1 is sending data to P_7 using the path $P_1 \rightarrow S_0 \rightarrow S_4 \rightarrow S_8 \rightarrow S_6 \rightarrow S_3 \rightarrow P_7$. Both routes utilize the link $S_0 \rightarrow S_4$ and both pass through S_6. Each switching element can handle multiple flows involving distinct links without loss of performance, but the competition for the link $S_0 \rightarrow S_4$ will result in either data loss* or reduced data injection rates on P_0 and P_1, ultimately leading to a loss of performance.

An alternative to static routing is *adaptive routing*, which dynamically changes the route between two endpoints to balance the traffic over all available links. The network hardware must be specifically designed to support adaptive routing, although some of the advantages of adaptive routing may also be gained with hardware that only supports source or static routing by using a technique called *dispersive routing*, in which data transmitted between a pair of endpoints travels along multiple, randomly selected paths.

* In this case the employed message-passing software will retransmit the data.

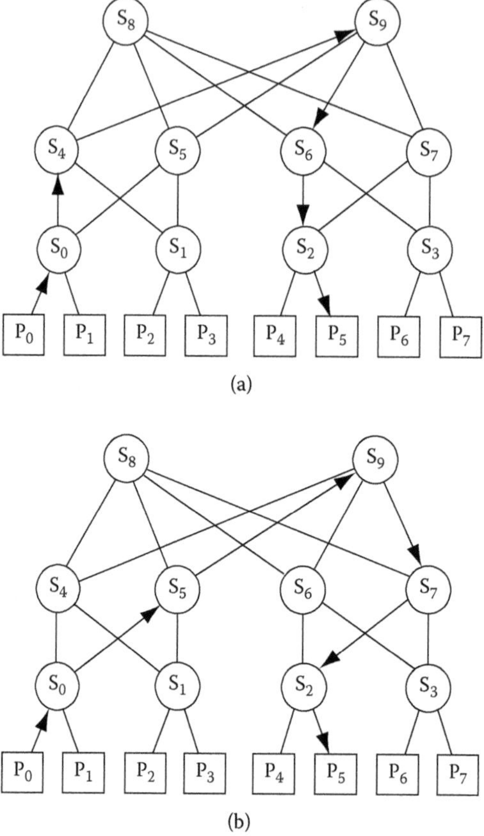

FIGURE 2.4
Two possible routes through a fat tree network for data sent from P_0 to P_5. In (a) the data travels along the path $P_0 \rightarrow S_0 \rightarrow S_4 \rightarrow S_9 \rightarrow S_6 \rightarrow S_2 \rightarrow P_5$. In (b) the data travels along the path $P_0 \rightarrow S_0 \rightarrow S_5 \rightarrow S_9 \rightarrow S_7 \rightarrow S_2 \rightarrow P_5$. Routes (a) and (b) will deliver equivalent performance in an otherwise idle network.

Example 2.1 Static versus Adaptive Routing
Figure 2.6 illustrates the performance on a Linux cluster[3] of two networking technologies using static and adaptive routing, respectively. The static routing data were collected using 4x InfiniBand™ hardware, and the adaptive routing data were obtained with 10 Gigabit Ethernet equipment employing a Woven Systems, Inc., 10 Gigabit Ethernet switch, which adaptively modifies the routing algorithm when overutilized links are detected. The adaptive routing scheme is capable of sustaining the same bandwidth as the number of nodes increases, but static routing leads to a significant performance drop when the number of nodes exceeds the number connected via a single switch, producing congestion in the network links.

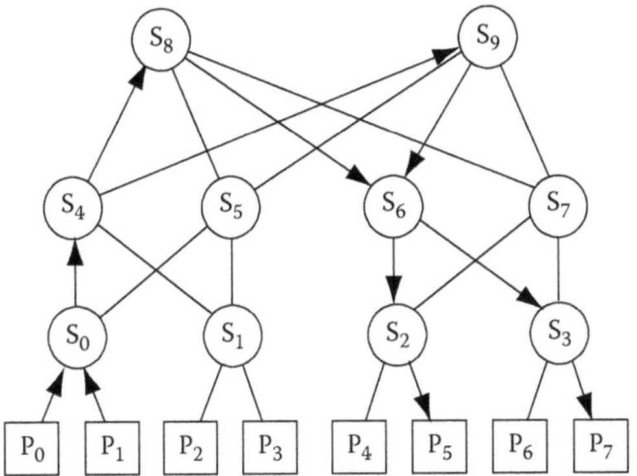

FIGURE 2.5
Network congestion caused by simultaneous data flows from $P_0 \rightarrow P_5$ and $P_1 \rightarrow P_7$. The link from $S_0 \rightarrow S_4$ must be used by both traffic flows, limiting network performance. The use of S_6 by both traffic flows will not cause congestion because switching elements can handle multiple flows involving distinct links without performance loss.

2.2.3 Network Performance

The application developer typically assumes that the network is flat and complete, that is, any given processing element has equal access to all others regardless of other network traffic. However, as we have seen in section 2.2.2,

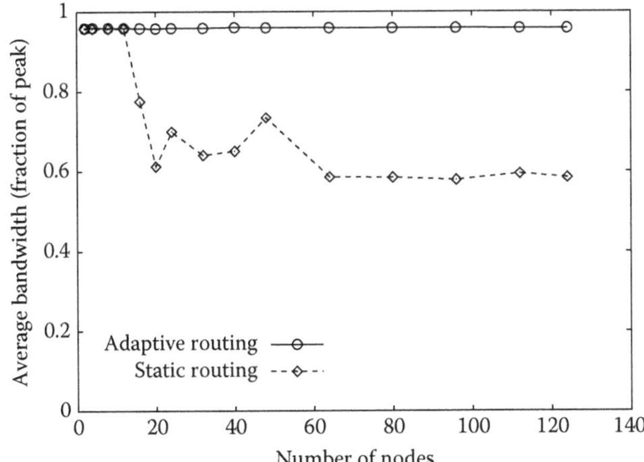

FIGURE 2.6
Average bandwidth obtained with adaptive and static routing, shown as a fraction of the peak bandwidth measured for each case. InfiniBand with 4x single data rate was used for static routing, and 10 Gigabit Ethernet was used for adaptive routing. Data were obtained on a Linux cluster[3] and represent average measured bandwidths for the Sandia Cbench Rotate benchmark.[12]

certain network traffic patterns can lead to congestion and result in less than ideal performance. We will here examine the performance of the network in the absence of congestion, and a measure for the amount of potential congestion will be discussed in section 2.2.4.

As a first approximation, after the first word[†] of data enters the network, the subsequent words are assumed to immediately follow as a steadily flowing stream of data. Thus, the time required to send data is the sum of the time needed for the first word of data to begin arriving at the destination (the latency, α) and the additional time that elapses until the last word arrives (the number of words times the inverse bandwidth, β):

$$t = \alpha + n_{\text{word}}\beta. \tag{2.1}$$

The latency can be decomposed into several contributions: the latency due to each endpoint in the communication, t_{endpoint}; the time needed to pass through each switching element, t_{sw}; and the time needed for the data to travel through the network links, t_{link} thus, the latency can be expressed as

$$\alpha \approx t_{\text{endpoint}} + n_{\text{sw}}t_{\text{sw}} + n_{\text{link}}t_{\text{link}} \tag{2.2}$$

where n_{sw} and n_{link} are the number of switches and links, respectively.

The endpoint latency, t_{endpoint}, consists of both hardware and software contributions and is on the order of 1 μs. The contribution due to each switching element, t_{sw}, is on the order of 100 ns. The t_{link} contribution is due to the finite signal speed in each of the n_{link} cables and is a function of the speed of light and the distance traveled; it is on the order of 1–10 ns per link. The total time required to send data through the network depends on the route taken by the data through the network; however, when we discuss performance modeling in chapter 5, we will use an idealized machine model where α and β are constants that are obtained by measuring the performance of the computer of interest.

The bandwidth of the network relative to the computational power of the nodes is another critical performance factor. The greater the computational power of the nodes, the greater is the need to rapidly transfer data between them; hence, greater computational power requires a higher bandwidth (that is, a smaller β). Let us look at the relationship required between the computational power and the bandwidth for efficient utilization of a parallel computer. The computational power of a processor can be expressed as $1/\gamma$, where γ is the time needed to perform a floating point operation (see section 5.3). The efficiency with which an application can use a parallel computer (defined in section 5.2) can be expressed as a function of α, β, γ, and the number of processors, p,

$$E(p, \alpha, \beta, \gamma) = \frac{1}{1 + f_\alpha(p)\alpha/\gamma + f_\beta(p)\beta/\gamma + \cdots} \leq 1 \tag{2.3}$$

[†] For our purposes, a word is defined to be eight bytes, since double precision data are usually used in quantum chemistry.

where $f_\alpha(p)$ and $f_\beta(p)$ are nonnegative application specific functions. A perfectly parallel application has an efficiency of one, and obtaining an efficiency as close to one as possible is desirable. Typically $f_\alpha(p) \ll f_\beta(p)$, and achieving high parallel efficiency therefore requires attention to the coefficient multiplying $f_\beta(p)$. A parallel architecture is said to be *balanced* if it is designed so that the γ to β ratio enables the machine to deliver adequate efficiencies for the machine's target applications.

2.2.4 Network Topology

The topology of the network strongly affects both its cost and its performance. The cost depends mostly on two parameters: the number of simple switching elements needed, n_{sw}, and the number of links supported by each switching element, known as the *degree* of the switch. The performance is also influenced largely by two topological factors, one of which is the number of switching elements that data must pass through to reach its destination, n_{hop}. The second factor determining the network performance is the bisection width, B_C, which is a measure for the network congestion that an application can generate. B_C is defined as the minimum number of network links that must be cut to completely separate the nodes into two sets of equal size. We here only consider bidirectional network links, that is, links on which data can be sent in both directions at the same time, and we will count each connection that is cut as two towards the computation of B_C. Topologies for which the bisection width equals the number of processors, $B_C = p$, are known as full bisection width networks.[‡] From a performance perspective, full bisection width networks are desirable because of the potential for low congestion in the network, even with the substantial traffic flows that are required by many high-performance computing applications. In practice, actual observed performance depends, among other factors, on the routing method used between the pairs of nodes, and even a full bisection width network can experience performance loss due to congestion. Nonetheless, B_C is a useful performance metric because it provides an upper bound on the number of links that can be utilized between node pairs in the worst-case bisection.

In the following, we will describe several network topologies, and Table 2.2 gives the values of the degree, n_{hop}, B_C, and n_{sw} for these topologies. Topologies of particular interest are the crossbar, which provides the building block for more complex networks, the mesh and torus networks, which provide a good balance of cost and performance that enables very large computers to be built, and the fat tree, providing excellent performance at low cost for intermediate numbers of nodes.

[‡] Bidirectional links are sometimes counted only as one link for the purpose of determining B_C, and a bisection width of $p/2$ then corresponds to a full bisection width network.

TABLE 2.2

Well-known network topologies and their performance characteristics. The network degree is the number of links supported by each simple switch, and p is the number of processors; n_{hop} is the maximum number of switches through which data must pass, and B_C and n_{sw} represent the bisection width and the number of simple switching elements, respectively

Network	Degree	n_{hop}	B_C	n_{sw}
Crossbar	p	1	p	1
Ring	2	$p/2+1$	4	p
Mesh	4	$2\sqrt{p}-1$	$2\sqrt{p}$	p
Torus	4	$\sqrt{p}+1$	$4\sqrt{p}$	p
3D Torus	6	$3/2p^{1/3}+1$	$4p^{2/3}$	p
Hypercube	$\log_2 p$	$\log_2 p+1$	p	p
k-ary Fat Tree	$2k$	$2\log_k p-1$	p	$(p/k)(\log_k p-1/2)$

2.2.4.1 *Crossbar*

The *crossbar* topology can directly connect any number of distinct pairs of processing elements. A crossbar connecting four nodes is shown in Figure 2.7 (a). The internal complexity of the crossbar is high, as is illustrated in Figure 2.7 (b). Because the degree of the switch in a crossbar network must be equal to the number of processing elements, p, it is costly to build crossbar switches for large p. Crossbars with low-degree switches, however, are frequently used as building blocks to construct larger networks for a number of different network topologies. Typically, the degree of a crossbar switch is limited by the

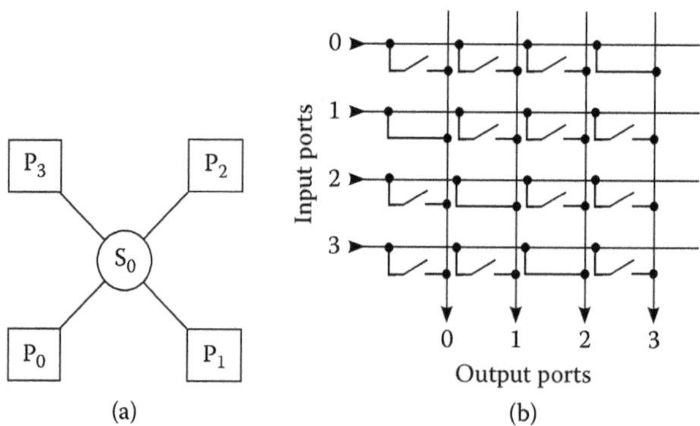

FIGURE 2.7
A crossbar network topology provides complete connectivity between all of the processing elements as shown in (a). This crossbar is a single degree–four switching element connecting four processing elements. The internal construction of a crossbar with four bidirectional ports is shown in (b). In the configuration shown, data from input ports 0, 1, 2, and 3 flow to output ports 3, 0, 1, and 2, respectively.

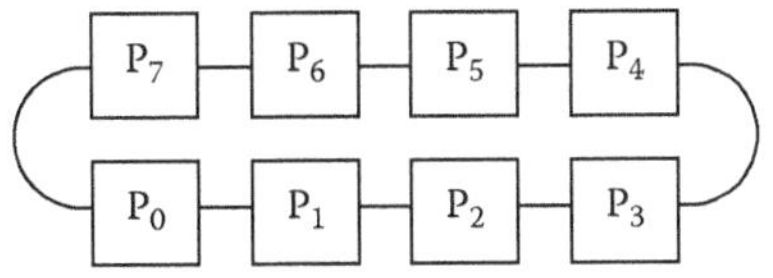

FIGURE 2.8
A ring topology directly connects each processing element with two neighboring processing elements.

functionality that can be fit into a single silicon chip. This chip must contain: a number of high-speed serializer/deserializers (SERDES) that convert data traveling on a wire to data held in the circuits on the chip; the crossbar; routing logic; buffers to store data while waiting for output links to become available; and a processor. An example is the Mellanox InfiniScale™ III switch chip, which has 96 SERDES, each moving data in both directions for a total for 8 Gbits/s. These SERDES are grouped into 24 groups of 4 or 8 groups of 12. A total of 768 Gbits/s of data can pass through the switch chip.

2.2.4.2 *Ring*

The *ring* topology, shown in Figure 2.8, connects each processing element to two neighbors. A ring is therefore inexpensive to build, but the ring topology also has a small bisection width, $B_C = 4$ (two bidirectional links are the minimum number of links that must be cut to separate the nodes into two halves). With a fixed bisection width, $B_C = 4$, compared with an ideal bisection width of p, the ring topology becomes increasingly inadequate as p increases. Hence, the ring topology is not used to connect a large number of processing elements, although rings can be employed to connect a smaller number of processors within a large-scale network (see Figure 2.15 for an example).

2.2.4.3 *Mesh and Torus*

The *mesh*[§] and *torus* topologies, as shown in Figure 2.9, place the switching elements in a regular grid with connections from each switch to its nearest neighbors. Torus networks are logically formed from meshes by connecting each switch on the boundary to the switch on the opposite boundary of the mesh. In practice, the long cable runs required to connect the switches on the opposite sides of a row can be avoided by doing the wrap-around in the torus as follows: connections are first established between all the odd-numbered switches in a row and then between the even-numbered switches, and these two sets are then connected at both ends of the row. A similar procedure can be used for the columns. Using this scheme, the cable lengths are roughly twice as long as in the mesh, but long cable runs are not required for the wrap-around.

[§] The term "mesh" is sometimes used to refer to what we in this book denote as an ad hoc grid.

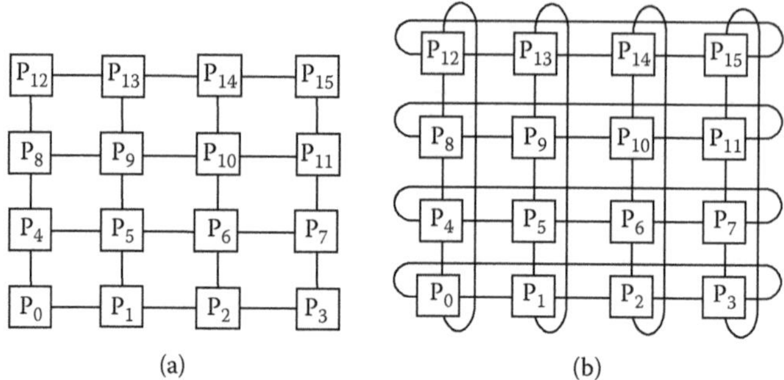

FIGURE 2.9
A two-dimensional mesh (a) and torus (b) network.

A torus network generally provides better performance than a mesh: in a torus, all nodes are equivalent, and n_{hop} is only about half that of the mesh while B_C is twice that of the mesh. The torus and mesh can also be generalized to three dimensions. The performance metrics listed in Table 2.2 for the mesh and torus are valid when the numbers of nodes in each dimension of the network are equal, although, typically, this condition is not met. To make it possible to build parallel computers of arbitrary size and to reduce manufacturing costs, mesh and torus networks are made from identical repeating units. Thus, it is natural to map the dimensions of the network to the physical dimensions of the computing facility, and mesh and torus networks typically do not have the same number of nodes along all dimensions. The values of B_C and n_{hop} in Table 2.2 are upper and lower bounds, respectively, for general mesh and torus networks.

2.2.4.4 Hypercube

In a *hypercube* topology, the number of nodes, p, is a power of two, $p = 2^k$, where k is the degree, or dimension, of the hypercube. A zero-dimensional hypercube is a single node, and a one-dimensional hypercube is a pair of nodes with a network link between them. A $(k+1)$-dimensional hypercube is constructed from two k-dimensional hypercubes by connecting each node with its counterpart in the other hypercube. A four-dimensional hypercube is shown in Figure 2.10. The hypercube topology was used for some of the early large-scale parallel computers, for instance, by Intel and nCUBE in the 1980s, but today the hypercube topology has largely been displaced by other topologies.

2.2.4.5 Fat Tree

A *fat tree* topology is shown in Figure 2.11. All the processing elements are at the lowest level of the tree, and above these the switching elements are placed in several levels, or stages. The first stage of switching elements connect downward to the processing elements and upward to the second stage of switching

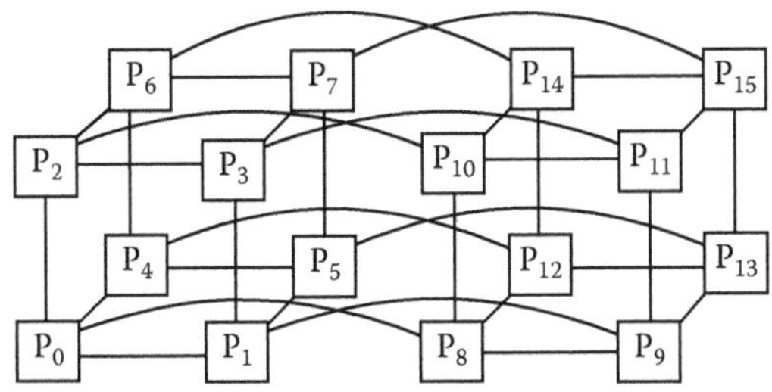

FIGURE 2.10
A hypercube network with degree $k = 4$.

elements; the second stage of switching elements connect downward and upward to the first and third stages of switching elements, respectively, and so forth. At each stage (except at the top), the total downward and upward bandwidths are equal, but the bandwidth going upward is concentrated into fewer, "fatter" links. The switches in the fat tree network are not all identical: as we move up the stages, the switches become more powerful.

Constructing a network from identical switches is much less costly, however, and in practice fat tree networks are usually constructed as shown in Figure 2.2. The fat tree network illustrated in Figure 2.2 is a binary fat tree, and it is also hierarchically constructed from multiple stages; however, instead of concentrating traffic to pass through fewer and more powerful switches as we move up the tree, the traffic is split between multiple switches. In general, a k-ary fat tree is constructed from switches with degree $2k$, and each switching element has k downward and k upward links except at the top, where there are $2k$ downward links. A k-ary fat tree with n stages can support up to $p = k^n$ nodes. The top stage only requires half as many switches as each of

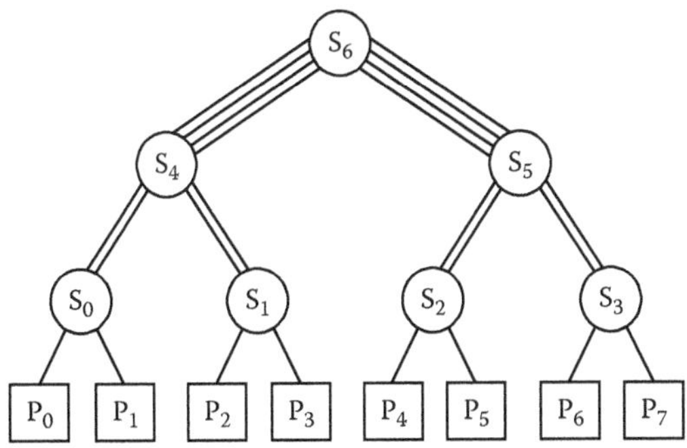

FIGURE 2.11
A fat tree network formed by progressively increasing the link bandwidth as we move up the tree: for each switch, except at the top level, the bandwidth on upwards links matches the sum of the bandwidths of the lower links.

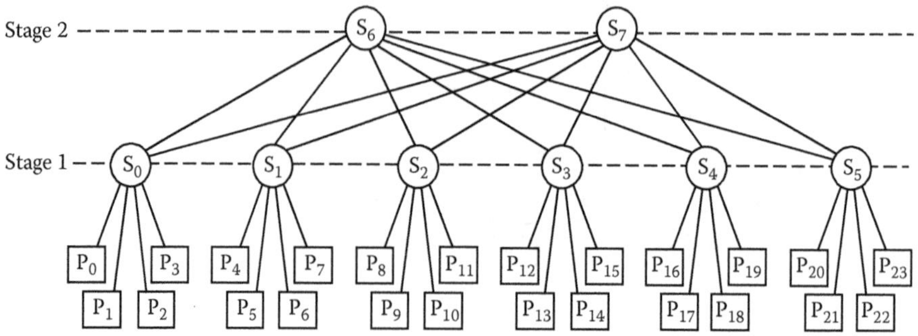

FIGURE 2.12
A two-stage, two-to-one oversubscribed fat tree network. Each stage 1 switch has four links to processing elements, but only two links to stage 2 switches.

the lower stages (when k is even) since the $2k$ links in each of the top switches are all directed downward. The fat tree is a full bisection width network, and implementations of fat trees using both adaptive routing and static routing exist.

The cost of a fat tree network can be reduced by making the tree oversubscribed. An oversubscribed fat tree has more downward links than upward links at one or more stages in the tree, and it has a reduced bisection width, B_C; if the bisection width has been reduced by a factor of a by oversubscription, that is, $B_C = p/a$, the tree is said to be a-to-one oversubscribed. A two-to-one oversubscribed fat tree is shown in Figure 2.12; in this network, the switches in the first stage are each connected to four nodes but have only two links to the switches in the second stage.

2.2.4.6 Bus

A *bus* network is illustrated in Figure 2.13. In bus networks, only one pair of ports can communicate at a time, and communication between the remaining ports is blocked until the network becomes available again. This behavior severely limits the ability of a bus network to support many processing elements. Bus networks were among the first networks deployed in inexpensive, modestly sized, parallel computers, and Ethernet hubs and Ethernet networks

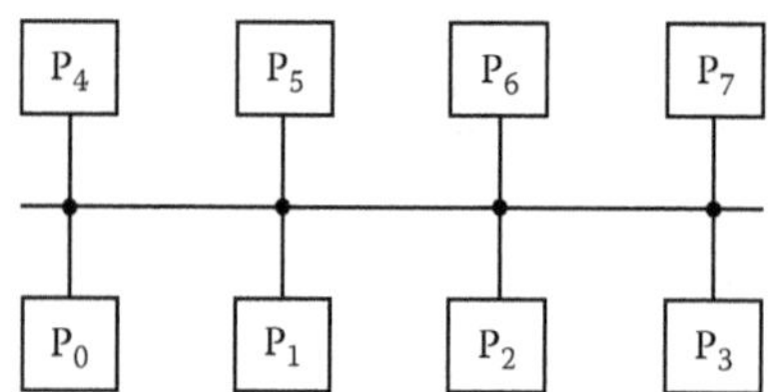

FIGURE 2.13
A bus network. Only a single pair of processing elements can communicate at any given time.

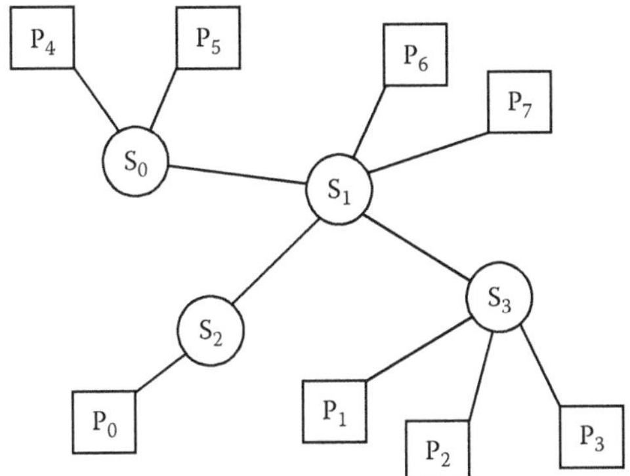

FIGURE 2.14
An ad hoc grid network. Ad hoc topologies are irregular, and the nodes can be geographically dispersed.

formed by tapping each computer into a single coaxial cable are examples of bus networks. Because of the availability of inexpensive crossbar components from which larger-scale networks can be constructed, bus networks are now uncommon.

2.2.4.7 Ad Hoc Grid

An *ad hoc grid* topology typically arises in computing environments that were not purpose-built to function as a single computer. The nodes in the grid are loosely coupled, as illustrated in Figure 2.14; a node may be administratively independent of the other nodes and may be distant as well, perhaps even connected over the Internet. The field of grid computing concerns itself with authentication, resource scheduling, data movement, and loosely-coupled computation in such environments. A grid network's performance, however, is too low (the bisection width is too small and the latency too large) to be of direct interest for the quantum chemistry applications discussed in this book.

2.3 Node Architecture

We have seen in the previous section that a massively parallel computer consists of a number of *nodes* connected via a communication network, and that the nodes comprise small groups of processors that share memory and other resources, although a node may also contain just a single processor. Each individual node in a parallel computer is typically essentially the same as a personal computer with the addition of specialized hardware (a *host channel adaptor* (HCA) or a *network interface card* (NIC)) to connect the computer to the network. Parallelizing an application is not strictly a matter of providing for parallelism across the nodes. Internally, each node has a significant amount of parallelism, and this introduces several additional layers of complexity to developers that seek to fully utilize the computer's resources. In this section we will discuss the logical internal structure of a node. The nodes in a parallel computer are composed of one or more SISD type processors, which execute instructions that may read and write memory locations, as needed. Modern processor designs take elaborate measures to hide the great discrepancy between memory speeds and processor speeds. These techniques include: caching data in high-speed memory nearer to the processor; processing instructions out-of-order so that instructions that do not have unfulfilled memory dependencies can continue ahead of other instructions that are waiting for data; and interleaving the processing of several instruction streams so that when one stream blocks, another stream can proceed. This latter approach is accomplished at the finest degree possible with *simultaneous multi-threading* (SMT), where individual instructions from multiple threads of execution can be selected for execution by the processor. These hardware-supported threads can correspond to either software-level threads or separate processes (see chapter 4 for more information on processes and threads).

Processors have evolved to become quite complex due to dramatic increases in the amount of functionality that can be placed on a single silicon chip. Initially, as the level of complexity started to increase, *instruction-level parallelism* (ILP) was introduced to allow instructions to be overlapped in execution by processing them in several stages forming a *pipeline*. The next advancement in ILP was the provision of multiple functional units in hardware, permitting instructions whose results are independent of each other to be executed simultaneously. At present, integration scales are so large that substantial additional improvements can no longer be achieved within a single processor, and therefore multiple processors are now packaged together on the same chip. The individual processors on such a chip are called *cores*[¶], and the resulting chip is referred to as a *multicore* chip. Cores on the same chip often share resources, in particular, a portion of their memory caches is typically shared. The integration level is expected to continue to rise in the future, resulting in the eventual appearance of *manycore* chips with a substantial number of cores on each chip.

Nodes often provide multiple sockets, each of which can accommodate a multicore chip (two or four sockets are common, and three-socket configurations exist as well). Figure 2.15 shows an example of a node architecture based on the quad-core AMD Opteron™ chip. Each chip has four cores (or processors, as we use the term here). Each core has its own first (L1) and second (L2) level caches. All four cores share the third level cache (L3). A crossbar switch connects the processors to two memory controllers for accessing off-chip memory. The crossbar has three additional high-speed links that can be used to connect other multicore chips or input/output devices.

¶ Often the multicore chips themselves are called "processors." We will use the term "processor" to refer to individual cores, unless the context clearly indicates otherwise.

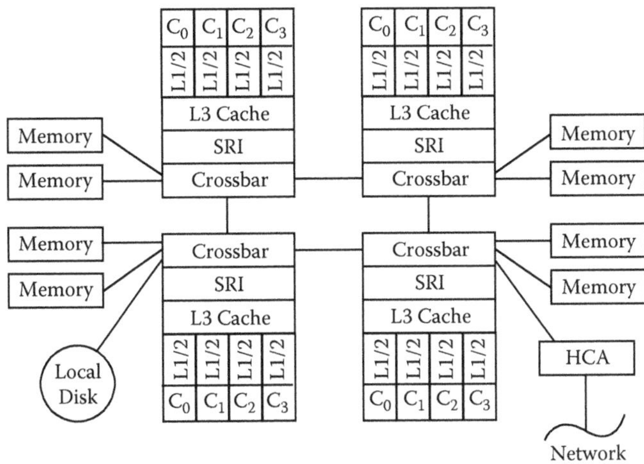

FIGURE 2.15
An example node consisting of four quad-core AMD Opteron chips. Memory, processors, and the parallel machine interconnect are reached though a ring network formed by connecting the crossbar switches. A single quad-core AMD Opteron chip provides four cores (C_0–C_3), two memory controllers, the System Request Interface (SRI), the crossbar, and the L1, L2, and shared L3 caches.

In Figure 2.15, a single HCA is connected, which provides connectivity to the other nodes on the network. The node shown has a *shared memory* architecture in which all the sixteen processors have direct access to all the memory. In a shared memory architecture, data from the same memory can be read by several processors and, hence, may be stored in multiple caches; if one of the processors then changes its data, the system may have two inconsistent copies of this data, resident in different cache entries. This situation is dealt with by *cache-coherency* support in hardware, which will invalidate data in other caches whenever the data in the local caches change. Such invalidation generates additional traffic on the network connecting the processors within the node, and this can impact the performance of the computer, especially when the number of processors in the node is large.

Shared memory computers in which all processors have equal access to all memory in the system are referred to as *symmetric multiprocessors* (SMP), and may also be called *uniform memory access* (UMA) computers. In the node shown in Figure 2.15, references to memory may need to pass through one, two, or three crossbar switches, depending on where the referenced memory is located. Thus, this node technically has a *nonuniform memory access* (NUMA) architecture, and, since the node is cache-coherent, this architecture is called ccNUMA. However, since the crossbar switches in the quad-core AMD Opteron implementation of ccNUMA exhibit high performance, this particular node would typically be considered to be an SMP.

Also shown in Figure 2.15 is a disk system. Often nodes do not have their own disks, but when present disks can be used as an additional means (much slower, but much less expensive, than memory) to store large amounts of data. Storage is discussed in more detail in section 2.4.2.

Modern processors, together with the operating systems that control them, provide multiple abstractions of the hardware within a node. *Virtual memory* and *multi-tasking* are two of the most important of these abstractions. Together, they allow multiple programs to run on the hardware, each appearing to have sole access to it. Each of these programs will run in its own *process*. Each process has virtual memory, which allows it to access memory as if no other processes were present. The hardware provides facilities that efficiently map addresses in the virtual memory address space into the node's physical memory. *Multi-tasking* refers to the ability to run multiple processes at once, with the processes alternately executing on the processor. This allows all processes to steadily make progress in their computations, even though there might be only a single processor. In multi-tasking, a process is periodically stopped and its state saved, another process' state is restored, and that process begins execution on the processor. This procedure is repeated for all processes. Together, multi-tasking and virtual memory create the illusion that each process has access to a dedicated processor, albeit one that may be slower and have less memory than the actual processor.

Each process can have multiple *threads* of execution. Threads provide independent execution contexts that share the same virtual memory. Nodes with multiple processors can schedule different threads and processes to execute concurrently. Sharing of data between processes is discussed in chapter 3, and an overview of programming with multiple threads is given in chapter 4.

Another abstraction of the hardware is the *virtual machine*. This has been available for decades in mainframe computers but only recently appeared in mainstream processors. The virtual machine abstraction takes virtual memory and multi-tasking a step further by more completely virtualizing the processor's capabilities. Virtual machines allow multiple operating systems to run concurrently.

2.4 MIMD System Architecture

Each of the individual nodes discussed in the previous section can be a MIMD parallel computer. Larger MIMD machines can be constructed by connecting many MIMD nodes via a high-performance network. While each node can have shared memory, memory is typically not shared between the nodes, at least not at the hardware level. Such machines are referred to as *distributed memory computers* or *clusters*, and in this section we consider such parallel computers in more detail.

2.4.1 Memory Hierarchy

Distributed memory computers are a challenge to program because of their deep hierarchical structure. We will examine their memory hierarchy by constructing a hypothetical computer using a node architecture similar to the one depicted in Figure 2.15, which contains four processing chips, each with its own local memory and four processor cores. In this hypothetical computer, our nodes are connected using the oversubscribed fat tree network topology shown in Figure 2.12. The resulting computer has a total of $4 \times 4 \times 24 = 384$ processors. The memory hierarchy is shown in Table 2.3. Different layers in this memory hierarchy are handled differently and to different degrees by various layers of software. At the lowest layer in the hierarchy are the processor registers. These hold a very small amount of data, which is immediately available for use by the processor. The compiler and very highly optimized math kernels perform the work of allocating registers to data. The next level up in the memory hierarchy is the memory cache, which itself is typically subdivided into several, progressively larger and slower, levels. Math kernels are optimized to access data in a way that keeps the most frequently used data in the fastest caches. Compilers can also take cache performance considerations into account as they optimize the executable code. Application programmers typically are not concerned with the details of the cache, but general programming principles such as striving to access data in loops with a stride of one are the result of attempting to make the best use of cache. The next level in the memory hierarchy is the main memory of the node, which provides storage that is much larger and much slower than the memory caches. Because, in our example, some memory is local to each chip, some is local to an adjacent chip, and some memory is two hops away, the memory can also be subdivided into several levels. It is typically up to the operating system to optimize access to main memory by allocating data for processes in the memory attached to the processor on which the process is running. The final levels in the memory hierarchy correspond to remote memory on other nodes. Because data transfered between two nodes may pass through either one or three switching elements in this example, remote memory can also be subdivided into two different levels. Applications usually directly manage the movement of data between nodes, often using message-passing techniques such as those discussed in chapter 3. However, applications typically use an idealized network model that ignores the number of switch hops between nodes. Getting good performance out of such deep memory hierarchies is a hard task, which will grow even more difficult as improvements in memory and network performance lag improvements in processing power.

2.4.2 Persistent Storage

There are two principal ways to provide persistent storage in a parallel computer: each node may have its own local disk, or a common storage device may be shared by all nodes and accessed via a network. The nodes in a parallel

TABLE 2.3

The memory hierarchy in a hypothetical MIMD machine constructed from the nodes in Figure 2.15 and the network in Figure 2.12. Logically equivalent data location types are subdivided by how many network hops are needed to reach them, n_{hop} (intra-node hops for local memory, inter-node hops for remote memory). The estimate of the time required to access memory at the given location is t_{access} (this time is hypothetical and not based on any particular hardware). The level of treatment typically used for each memory level is given for the several layers of software

			Typical Level of Treatment			
Data Location	n_{hop}	t_{access} (μs)	**Compiler**	**Operating System**	**Math Kernel**	**Application**
Register		0.0005	Maps data to registers.		Optimizes register and multilevel cache use.	Node storage (registers, cache, and local memory) considered uniform.
L1 cache		0.0015	Cache optimizations where possible.	Attempts to run processes near data for both cache and local memory.		
L2 cache		0.0060				
L3 cache		>0.0060				
Local memory	1	0.070				
Local memory	2	0.080				
Local memory	3	0.090				
Remote memory	1	1.5				Explicitly managed.
Remote memory	3	2.1				

computer are often diskless, and in this case all storage is networked. In computers that have local disks, network storage is usually provided as well as a way of making, for instance, home directories readily available on all nodes.

2.4.2.1 Local Storage

The fastest currently available disk drives have a bandwidth of around 100 Mbyte/s and a latency of about 3 ms. Multiple drives can be combined together as a *redundant array of inexpensive disks* (RAID) system to provide better bandwidth, but the latency cannot be improved in this way. Because disk access is considerably slower than memory access, the use of disk resources in a parallel computer must be carefully planned to avoid large performance penalties. For example, in applications involving storage of large quantities of data on disk, data compression can be employed to reduce the bandwidth required for retrieval of the data. An example of the use of this strategy in a quantum chemistry application is the compression of the two-electron integrals stored on disk in a parallel Hartree–Fock program.[4] In general, the use of local disk-based storage is not a parallel programming issue, and we will not consider this topic any further in this text.

2.4.2.2 Network Storage

Some form of network storage is typically available on parallel computers. Network storage can be provided by a single, special-purpose node, for instance, a server providing shared storage through the Network File System (NFS) protocol. While multiple nodes can access data concurrently on such a file server, the performance of the file server severely degrades as more nodes attempt simultaneous access. Alternatively, the parallel computer can be attached to a storage system that is essentially another parallel computer dedicated to providing high-performance storage. This kind of storage system has multiple high-speed connections to the parallel computer and can provide much higher performance. Lustre®[5] is an example of such a parallel file system.

Even though parallel disk storage systems can provide much higher bandwidth than either a local disk or network storage provided through a single server, the available performance typically is not sufficient to meet the needs of many quantum chemistry applications. Considering, for example, Fock matrix formation (discussed in chapter 8), an implementation may use either a direct method that computes the two electron atomic orbital integrals as needed, or a disk-based method that stores the integrals on disk and reads them in each iteration. The time required to compute an integral on a current processor is about 0.5 μs. On a large-scale computer with 20,000 processors, an integral would then be computed on average every 25 ps. If these integrals were read from disk rather than computed, matching this computation rate would require a bandwidth of 320 Gbytes/s—and even computers of this size usually cannot deliver such bandwidth. For smaller parallel computers, the storage systems are also unlikely to be able to deliver the necessary bandwidth: for example, using eight-processor nodes, the required per-node bandwidth would be 128 Mbytes/s/node, and sustaining this bandwidth to all nodes simultaneously is beyond the capabilities of the storage systems commonly found in clusters. These constraints limit the usefulness of network storage for quantum chemistry applications. Some quantum chemistry applications do make use of network storage, however, to allow reuse of data that are very expensive to compute. For example, parallel storage systems have been used to store the molecular orbital integrals in parallel implementations of correlated electronic structure methods such as the MP2-R12 methods.[6]

2.4.2.3 Trends in Storage

Being mechanical devices, disk drives will always have severe latency issues. Already, disk drives are being replaced in consumer devices by *NAND flash* memory, a solid state persistent storage medium. This memory has the advantage of a significantly lower latency and higher bandwidth than disk drives. It is currently more expensive per unit storage than disks, but the large manufacturing volumes needed to meet the demand for consumer devices using NAND flash are driving its price down, and NAND flash will find increasing use in computing. The availability of solid state storage in parallel computers would make the use of persistent storage desirable for a much wider range of methods than those that can use current disk technology.

2.4.3 Reliability

Many components must work in harmony for an application to successfully run on a large parallel computer, and this requires an exceptional level of reliability for all components. Although a personal computer that failed only once every five years would probably be very satisfactory, a 10,000 node computer built from nodes with that same individual failure rate would fail, on average, about every 4.4 hours. This figure is known as the *mean time between failures* (MTBF) for the computer and is the reciprocal of the rate of failure. Let us look at a collection of N types of components with n_i components of each type, and let each component type have a mean time between failures of MTBF_i. For this system, the overall mean time between failures will be the reciprocal of the sum of the rates of failure for the individual components

$$\mathrm{MTBF}_{\mathrm{system}} = \frac{1}{\left(\sum_{i=1}^{i\leq N} \frac{n_i}{\mathrm{MTBF}_i}\right)} \leq \min_{1\leq i\leq N} \frac{\mathrm{MTBF}_i}{n_i}. \tag{2.4}$$

Thus, the mean time between failures for the system is bounded by the weakest component type in the system, namely the one with the smallest $\frac{\mathrm{MTBF}_i}{n_i}$ value; for systems with many types of components, however, the mean time between failures is typically much smaller than that for the weakest component. If a particular type of component has a small MTBF, then overall system reliability can be increased by eliminating as many of these components as possible. Thus, nodes in a parallel computer often have no disk drive, because disk drives are one of the most failure-prone components. Generally, it is difficult to obtain an accurate mean time between failures for a system because manufacturers often overestimate the MTBF for components; also, MTBF claims made by manufacturers of parallel computers must be critically evaluated, as MTBF values are often miscalculated.

One way to mitigate the effects of failures is to introduce redundancy into the system for the components that are most prone to failure. Typically, systems with redundancy are designed so that when a single component fails, that component can be replaced without interrupting the operation of the system. The system will fail only when more than one failure occurs before the first failure can be repaired. Redundancy is commonly used for power supplies and fans. Networks such as the oversubscribed fat tree shown in Figure 2.12 have natural redundancy in the links between two switching elements. If one element fails, traffic can be routed around the failed link, albeit at reduced performance.

2.4.4 Homogeneity and Heterogeneity

A parallel computer is said to be *homogeneous* if the same program running on different nodes will produce exactly the same results in the same amount of time. Otherwise, the computer is considered to be *heterogeneous*. Most large-scale parallel computers today are built to be homogeneous. Ad hoc collections of nodes, and homogeneous systems that have had new hardware added after the original installation, are often heterogeneous.

Heterogeneity in a parallel computer can sometimes lead to reduced performance, incorrect results, or deadlock. Consider, for example, a parallel computer in which all nodes are identical, except for one node, which has a processor with a clock speed that is only half that of the processors on the other nodes. If a program running on this parallel computer distributes the computation by assigning the same amount of work to each node, the slower node will require twice as much time as the other nodes to complete its work. Hence, the overall time to complete the application will be twice as long as if all nodes had the same, faster, clock speed; approximately half of the computing power of the entire parallel machine could thus be lost due to the one slow node.

Heterogeneity in the employed system libraries or compilers can result in inconsistent or incorrect results for a parallel program, even when the hardware and executables are homogeneous. For example, using different dynamically linked system libraries (such as math libraries) on different nodes can lead to slightly different results for a program running on different nodes. Using different compilers or compiler flags can also change the results of floating point computations. For example, the Intel® architecture stores double precision floating point numbers using 80 bits in registers but only 64 bits in memory, and storing a floating point register to memory results in rounding. When a subroutine requires enough temporary data that the compiler cannot keep all of it in the processor's registers, the compiler must chose which data to store in memory, and this choice will affect the numerical result.

Programs that require that a certain routine, for a given set of input data, will produce bit-wise identical answers on all nodes employed may experience difficulties when run in a heterogeneous environment. For example, consider an iterative algorithm that uses the norm of a residual vector to determine if the calculation has converged and the iterative procedure can be exited. Slight numerical differences can cause the computation on different nodes to exit the iterative procedure after different numbers of iterations, and if inter-node communication is required to complete each iteration, the nodes that did not exit the iterative procedure would be left waiting for the other nodes that did exit, resulting in deadlock.

Note that even computers intended to be homogeneous, consisting of identical nodes and using the same software on every node, may not always, in practice, meet the conditions for homogeneity. Thus, factors beyond the control of the builder and user of the computer may cause otherwise identical nodes to run at different speeds, effectively creating a heterogeneous environment. Examples include: the slow-down of execution on a node caused by error correction performed by *error-correcting code* (ECC) that is employed to improve robustness of memory (if one bit is wrong, the error will be corrected, and execution will proceed); the reduction of a processor's clock speed to prevent overheating; and the malfunctioning and off-lining of one or more cores in a multicore chip, which reduces the computational power of the node but does not cause failure of the chip.

2.4.5 Commodity versus Custom Computers

Parallel computers broadly fall into one of two categories: *commodity clusters*, which are manufactured from commonly available components, and *custom high-performance computers* containing parts designed for, and used only by, a particular product line of computers. Commodity clusters usually have a lower cost because they use widely available parts that are produced in large volumes. They nearly always run an operating system that is an open-source variant of UNIX™, such as GNU/Linux, and are then also referred to as *Beowulf clusters*. Custom computers, however, typically support larger numbers of processors and provide higher levels of performance for applications that run well at such large scales.

Originally, custom parallel computers employed both custom processors and a custom interconnect, but today many custom computers use commodity processors combined with purpose-built interconnects. In our discussion of the mean time between failures (MTBF), we saw that a high MTBF in a large-scale computer requires very high reliability of the individual components, and large-scale custom computers are designed with this in mind. For instance, custom computers use fewer, more reliable fans, and some computers use only a single, large, and very reliable fan for an entire rack of nodes. Additionally, these computers are designed with a high level of redundancy for power supplies and do not use disk drives for individual nodes, as both of these components are also failure prone.

Example 2.2 The Sandia Red Storm Supercomputer
This is the computer that formed the basis for the Cray XT3™ supercomputer. Every node has a 2.4 GHz dual core AMD Opteron™ processor, and a direct network provides each node with a 6.4 GB/s bidirectional link to its own switch using AMD HyperTransport™. The switches connect to their six neighbors with 7.8 GB/s bidirectional links. A $27 \times 20 \times 24$ 3D mesh topology is used for a total of 12,960 compute nodes. An additional 320 nodes are used for I/O and service functions on each of the two networks to which the machine can be connected. The network latency, including software overhead, between nearest neighbors is 4.8 μs, and the worst-case network latency is 7.8 μs. Red Storm's parallel filesystem can maintain a rate of 50 Gbytes/s. An operating system specifically designed for high-performance computing (a *lightweight kernel*), named Catamount, is used on the compute nodes, and GNU/Linux runs on the service nodes. The High-Performance Linpack benchmark[7] (HPL) on this computer achieved 101.4 teraflops on 26,544 processors,[‖] placing it third on the June 2007 TOP500 list.[8]

Although, at any given time, the largest custom computers are faster than the largest commodity computers, the performance of commodity parallel computers has been improving steadily. This is illustrated in Figure 2.16, which depicts the HPL benchmark results from the TOP500 list for both the fastest custom and the fastest commodity parallel computers. An analysis of this kind is fraught with limitations: the TOP500 list is self-selected, the HPL benchmark may not be the most meaningful measurement of machine performance, and it can be difficult to assign certain computers to one of the commodity and custom categories. Nonetheless, some useful information can still be collected from this exercise. It is readily apparent that the performance of both commodity and custom parallel computers is exponentially increasing. The performance of custom computers has doubled about every twelve months from 1993 to 2007. Commodity computers started out more than thirty times slower than custom computers but improved their performance at a higher rate, doubling every nine months from 1995 to 2001. In 2002 the performance of commodity computers took a leap and almost reached the performance of custom computers but then started growing more slowly, doubling every sixteen months. Throughout the time period considered here, the performance of commodity computers has lagged that of custom computers by two to five years.

Parallel computers may also be distinguished based on whether they are allocated for *capacity* or *capability* computing, and this distinction is related to the commodity versus custom classification. Capacity computers are designed to obtain good throughput and high cost-effectiveness for running many small- to modest-sized jobs. Capability computers are reserved for

[‖] The number of processors exceeded $2 \times 12{,}960$ because some I/O nodes were also used for computation for this benchmark.

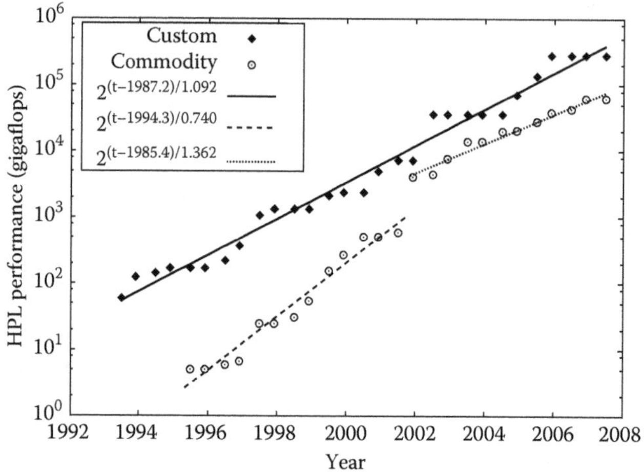

FIGURE 2.16
Semilogarithmic plot of the HPL benchmark value for the fastest custom and commodity computers, as reported in the TOP500 list.[13] Custom computers show exponential growth throughout this time period, doubling in performance about every year. The performance of commodity computers also improves exponentially, doubling every nine months from 1995 to 2001 and then slowing down, doubling about every sixteen months from 2001 to the present. The anticipated appearance of petaflop computers in 2008 will be in accord with the current growth trend of custom computers.

running the very largest of jobs that cannot run elsewhere due to their large requirement for processing power or memory. Commodity clusters are typically used for capacity computing, and custom computers are usually reserved for capability computing.

Example 2.3 The Sandia Thunderbird Cluster
This computer is a commodity cluster with 4,480 nodes. Each node has two 3.6 GHz single-core Intel® Xeon® processors. The interconnect is 4x single data rate InfiniBand, which provides a peak bidirectional data rate of 2 GB/s. The network topology is a 2-to-1 oversubscribed fat tree, and the operating system is GNU/Linux. This cluster ran the High-Performance Linpack benchmark[7] at a rate of 53 teraflops, placing it at number 11 on the June 2007 TOP500 list.[8]

2.5 Further Reading

An in-depth, yet very approachable, discussion of network architecture is presented in *Principles and Practices of Interconnection Networks* by Dally and Towles.[9] An up-to-date and informative discussion of node architecture can be found in the fourth edition of Hennessy and Patterson's *Computer Architecture: A Quantitative Approach,*[10] which also includes a CDROM with an appendix providing an overview of network architecture. A more detailed discussion of the dangers of using heterogeneous computers has been given by Blackford et al.[11]

References

1. Flynn, M. J. Some computer organizations and their effectiveness. *IEEE Trans. Comp.* C-21:948–960, 1972.
2. Hillis, W. D. *The Connection Machine.* Cambridge: MIT Press, 1989.
3. A Linux® cluster consisting of nodes with two single-core 3.6 GHz Intel® Xeon® processors (each with 2 MiB of L2 cache) connected via two networks: (1) a 10 Gigabit Ethernet network using Chelsio R310E iWARP host adaptors and a Woven Systems Inc. EFX 1000 switch; and (2) a 4x Single Data Rate InfiniBand™ network using Mellanox Technologies MT25208 InfiniHost™ III Ex host adaptors and a Topspin 540 switch. RDMA protocols were used for both InfiniBand and 10 Gigabit Ethernet, and all results were measured with the host adaptors in 8x PCI Express slots.
4. Mitin, A. V., J. Baker, K. Wolinski, and P. Pulay. Parallel stored-integral and semidirect Hartree-Fock and DFT methods with data compression. *J. Comp. Chem.* 24:154–160, 2003.
5. Braam, P. J. Lustre File System, a white paper from Cluster File Systems, Inc., version 2, 2007. http://www.clusterfs.com/resources.html.
6. Valeev, E. F., and C. L. Janssen. Second-order Møller–Plesset theory with linear R12 terms (MP2-R12) revisited: Auxiliary basis set method and massively parallel implementation. *J. Chem. Phys.* 121:1214–1227, 2004.
7. Petitet, A., R. C. Whaley, J. Dongarra, and A. Cleary. HPL—a portable implementation of the high-performance Linpack benchmark for distributed-memory computers. http://www.netlib.org/benchmark/hpl/.
8. Meuer, H., E. Strohmaier, J. Dongarra, and H. D. Simon. TOP500 supercomputing sites. June 2007. http://www.top500.org.
9. Dally, W. J., and B. Towles. *Principles and Practices of Interconnection Networks.* San Francisco: Morgan Kaufmann, 2004.
10. Hennessy, J. L., and D. A. Patterson. *Computer Architecture: A Quantitative Approach,* 4th edition. San Francisco: Morgan Kaufmann, 2007.
11. Blackford, S., A. Cleary, J. Demmel, I. Dhillon, J. Dongarra, S. Hammarling, A. Petitet, H. Ren, K. Stanley, and R. C. Whaley. Practical experience in the dangers of heterogeneous computing. In *Lecture Notes in Computer Science* 1184:57–64. Berlin: Springer, 1996.
12. More information on the Cbench benchmarks can be found on the World Wide Web at http://cbench-sf.sourceforge.net.
13. Meuer, H., E. Strohmaier, J. Dongarra, and H. D. Simon. TOP500 Supercomputing Sites. June 1993 to June 2007. http://www.top500.org.

3

Communication via Message-Passing

On distributed memory computers, the memory of a node is directly accessible only to the processes running on that node, and message-passing is the primary means for exchanging data between processes running on different nodes. Most parallel algorithms for distributed memory computers therefore use a programming model, the message-passing paradigm, in which processes exchange data by explicitly sending and receiving messages. Except for trivially parallel computational problems in which the processes can work completely independently of each other, parallel applications involve some data exchange between processes, and often a significant amount of communication is required. For instance, data needed by a process may have to be retrieved from a remote memory location associated with another process, or a manager process in charge of distributing and scheduling work will need to communicate with other processes to do so. The use of message-passing in a parallel program may affect the parallel performance in several ways. For example, the *communication overhead* (the time required to perform the communication) may be significant and may grow as the number of processes increases, and the presence of communication steps may require synchronization of all processes, which forces some processes to be idle while waiting for other processes to catch up.

Communication operations can be classified into three categories: *point-to-point* communication, which requires cooperation of the sending and receiving process and is the most basic type of message-passing; *collective* communication involving a group of processes; and *one-sided* communication, which enables one process to control the exchange of data with another process. In this chapter, we will discuss these types of communication and show examples of their application. While much of the discussion will not be specific to a particular message-passing library, we will use the Message-Passing Interface (MPI) to illustrate the concepts. MPI is the most widely used message-passing library in scientific computing, and it is available on most parallel platforms employed in this field. A brief introduction to MPI is provided in Appendix A.

3.1 Point-to-Point Communication Operations

Point-to-point communication, also referred to as pairwise communication, entails the sending of a message from one process to one other process. To accomplish this type of operation, an explicit call to a sending operation must be placed by the sending process while the posting of a corresponding receive operation is required by the receiving process. Point-to-point communication operations may be broadly characterized as either non-blocking or blocking, depending on whether execution control is blocked until the message transmission is complete. Application of point-to-point communication operations creates the possibility for writing *unsafe* parallel programs, including programs that produce a nondeterministic result due to race conditions and programs that cause deadlocked processes. It is important never to rely on any unsafe programming practices whatsoever in parallel programming; although an unsafe program may work reliably on one parallel architecture or using a specific version of a message-passing library, unsafe programs are very likely to cause problems, for example, when porting code to other computers, updating message-passing libraries, or using the application with new input data sets. In the following we will discuss blocking and non-blocking message-passing in more detail, illustrate their use in a few examples, and briefly address the use of safe programming practices that avoid deadlock and race conditions.

3.1.1 Blocking Point-to-Point Operations

A send or receive operation is said to be *blocking* if execution is blocked until the operation is complete. Completion of the send or receive operation is defined in terms of the associated buffer: an operation is complete when it is safe to reuse the buffer used for the message transfer. Thus, when a process posts a blocking send, execution for that process will be suspended until the send buffer can be safely overwritten. Likewise, after posting a blocking receive, execution will resume only when the data to be received is guaranteed to have been put into the receive buffer.

Using blocking send and receive operations in a program can reduce the memory requirement because the message buffers are guaranteed to be safe for reuse upon return of the functions so that no extra buffer space needs to be allocated for the message transfer (although the message-passing library may allocate the storage anyway). The use of blocking operations, however, creates the possibility for a *deadlock* situation. Deadlock arises when two or more processes are stuck waiting for each other. For example, one process may be waiting for a response from a second process, which, in turn, is waiting for the first process to respond. In Figure 3.1, we show a function that employs MPI blocking send and receive operations to send data between processes in a virtual ring in a manner that creates deadlock of all processes. In this case, every process first posts a blocking receive operation to receive a message from

```
void send_data_in_ring_deadlock(int *indata, int *outdata,
                               int ndata, int this_proc, int p)
{
  /* this_proc: process ID; p: number of processes */

  int next_proc;        /* send data to this process */
  int previous_proc;    /* receive data from this process */
  MPI_Status status;    /* Required by MPI_Recv */

  if (this_proc == p-1) next_proc = 0;
  else next_proc = this_proc+1;

  if (this_proc == 0) previous_proc = p-1;
  else previous_proc = this_proc-1;

  MPI_Recv(indata, ndata, MPI_INT, previous_proc, 0,
           MPI_COMM_WORLD, &status);
  MPI_Send(outdata, ndata, MPI_INT, next_proc, 0,
           MPI_COMM_WORLD);

  return;
}
```

FIGURE 3.1
A function (written in C) using MPI blocking send and receive operations and causing a deadlock of all processes. The function attempts to send data around in a virtual ring of p processes by letting each process, with process identifier `this_proc`, receive data from process `this_proc`−1 while sending data to `this_proc`+1 (process $p-1$ sends to process 0 to complete the ring). The blocking receive, `MPI_Recv`, (posted by all processes) never returns, however, because every process is waiting for the matching send, `MPI_Send`, to be called at the sending end. The employed MPI functions are explained in Appendix A.

another process, but no message is ever received because the process that is to send the message will be waiting to receive a message from another process. In general, deadlock can be avoided by using non-blocking communication operations.

3.1.2 Non-Blocking Point-to-Point Operations

A *non-blocking* send or receive operation will return control to the calling process immediately after the call has been posted. For a non-blocking send operation, therefore, control is returned to the calling process without checking whether the data has been copied out of the send buffer or transmission of the data started. A non-blocking receive, likewise, returns without determining whether the data to be received has been copied into the receive buffer. Non-blocking communication operations are thus an example of *asynchronous* operations, namely, operations that will return without requiring cooperation from a remote process. *Synchronous* communication operations,

```
void send_data_in_ring(int *indata, int *outdata,
                       int ndata, int this_proc, int p)
{
  /* this_proc: process ID; p: number of processes */

  int next_proc;          /* send data to this process */
  int previous_proc;      /* receive data from this process */
  MPI_Request request; /* Required by MPI_Irecv, MPI_Wait */
  MPI_Status status;      /* Required by MPI_Wait */

  next_proc = (this_proc == p-1 ? 0 : this_proc+1);
  previous_proc = (this_proc == 0 ? p-1 : this_proc-1);

  MPI_Irecv(indata, ndata, MPI_INT, previous_proc, 0,
            MPI_COMM_WORLD, &request);
  MPI_Send(outdata, ndata, MPI_INT, next_proc, 0,
           MPI_COMM_WORLD);
  MPI_Wait(&request, &status);

  return;
}
```

FIGURE 3.2
A modified version of the C function in Figure 3.1 using an MPI non-blocking receive, `MPI_Irecv`, to avoid deadlock. The `MPI_Irecv` returns without waiting for any message to arrive, and each process then proceeds to post the sending operation, `MPI_Send`. The `MPI_Wait` causes each process to wait at this point until the message to be received by `MPI_Irecv` has arrived. The employed MPI functions are explained in Appendix A.

on the other hand, require a function call on both the sending and receiving process before returning. When using non-blocking operations it is usually necessary at some point to ascertain that it is safe to overwrite the send buffer, or that the data has arrived in the receive buffer. This can be accomplished by calling a function (for example, `MPI_Wait`) that blocks until the previously posted non-blocking send or receive operation has completed, or by calling a function (such as `MPI_Test`) that tests for completion of the preceding send or receive but does not block. Attempting to read the receive buffer after posting a non-blocking receive can lead to a *race condition*, which is a non-deterministic program behavior caused by critical dependence on the relative timings of events. Thus, reading from the message buffer before ensuring that the receive has completed will produce a different result depending on whether the receive did already complete. Likewise, a race condition can be caused by trying to modify the send buffer after calling a non-blocking send operation.

In Figure 3.2, we show a modified version of the function from Figure 3.1, which sends data around in a virtual ring. In the modified version, each process first posts a non-blocking receive operation to receive data from another process. This receive function will return immediately without waiting for a message to arrive, and all processes can therefore proceed and post the subsequent send operations, sending out the data to be received by the non-blocking receives. Deadlock can thus be avoided by replacing the blocking receive used in Figure 3.1 with a non-blocking receive. Before returning, however, each process must wait for the preceding non-blocking receive to finish receiving its message (by calling `MPI_Wait`) so that the program will not exit before all the data has arrived.

The use of non-blocking point-to-point communication offers the potential for improving performance by overlapping communication with computation. A call to a non-blocking send or receive operation only briefly interrupts execution while the call is being posted, and the calling process can resume computation without waiting for the send or receive to complete. This also means that if a process initiates communication with another process that is not yet ready to respond to the message, the issuing process does not have to remain idle while waiting for the response. However, overwriting the send buffer or using the data in the receive buffer must not be attempted until the non-blocking communication operation is complete. The extent to which it is possible to overlap communication and computation depends on the actual implementation of the non-blocking communication operations. Rendezvous protocols, which may require an initial message exchange (a "handshake") between the sending and receiving processes before the data transfer is initiated, are often used for non-blocking message-passing to avoid creating extra message copies. The requirement of such a handshake, however, can reduce the ability to overlap communication and computation.[1] Newly developed protocols for non-blocking message-passing hold promise for achieving a better overlap of communication and computation and, hence, improving parallel efficiency and scalability for programs employing non-blocking operations.[1]

3.2 Collective Communication Operations

Collective communication operations are message-passing routines in which a group of processes are involved simultaneously. If all processes in a parallel application are involved, the operation is also referred to as *global communication*. Collective communication operations implemented in message-passing libraries such as MPI offer a convenient way to exchange data between processes or perform other operations that involve a group of processes. Collective operations that simply move data between processes include: broadcasts, in which a process sends a message to all other processes; scatter operations involving a process sending a different message to all other processes; and gather operations, the reverse of scatter operations, gathering data from all processes onto one process. Another type of collective operations are reduction operations, which perform data movement as well as some computations on the data. Finally, a third category of collective communication operations serve to synchronize processes, for instance, by introducing a barrier into the program beyond which no process can pass until it has been reached by all processes. In Appendix A we list examples of commonly employed MPI collective communication operations and illustrate their use.

Collective operations are relatively easy to use, and they eliminate the need for the programmer to develop custom-designed and potentially complicated parallel communication schemes. Additionally, efficient algorithms have been implemented for many collective communication operations on a variety of parallel platforms, increasing the portability of parallel code that employs these operations. Potential disadvantages of collective communication include the introduction of communication bottlenecks and reduced efficiency resulting from process synchronization. In the following we will discuss a few of the most commonly used collective communication operations, present algorithms for carrying out these operations, and analyze their cost for the purpose of performance modeling. Examples of algorithms employing collective communication are given in sections 8.3 and 9.3 and in chapter 10.

3.2.1 One-to-All Broadcast

A one-to-all broadcast is a communication operation in which data from one process is sent to all other processes. The data distribution before and after the broadcast is illustrated in Figure 3.3. A commonly used algorithm for performing the one-to-all broadcast is the binomial tree algorithm, which is illustrated in Figure 3.4 for a case with eight processes labeled P_0–P_7. In the first step, P_0 sends its data to P_4; in the second step, P_0 and P_4 send their data to P_2 and P_6, respectively; and in the final step P_0, P_2, P_4, and P_6 each send their data to the process whose identifier is one higher than that of the sending process. In the general case, assuming we have p processes where p is a power of two, the ith step in the broadcast involves 2^{i-1} processes each sending data to the process whose process identifier is $p/2^i$ higher. The entire broadcast process requires $\log_2 p$ steps, and in each step data of length l is

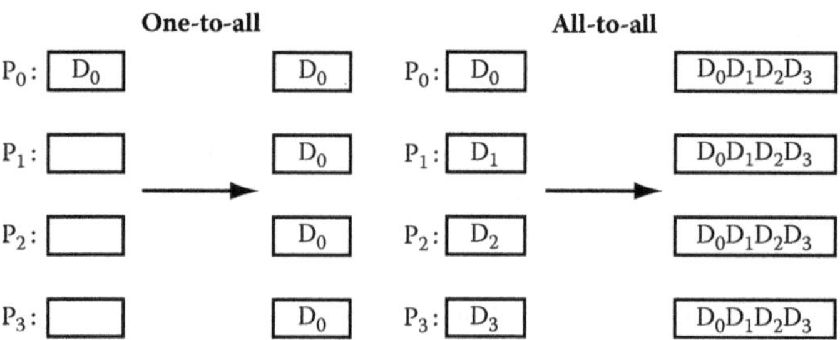

FIGURE 3.3
Data distribution before and after one-to-all and all-to-all broadcast involving four processes, P_0–P_3. D_i represents the data initially associated with process P_i.

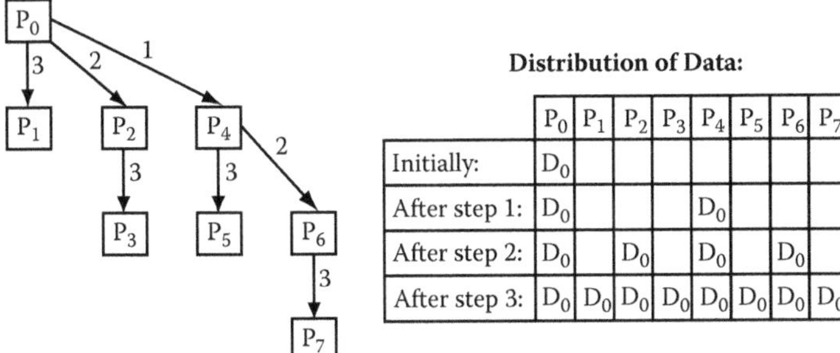

	P_0	P_1	P_2	P_3	P_4	P_5	P_6	P_7
Initially:	D_0							
After step 1:	D_0				D_0			
After step 2:	D_0		D_0		D_0		D_0	
After step 3:	D_0	D_0	D_0	D_0	D_0	D_0	D_0	D_0

FIGURE 3.4
One-to-all broadcast using a binomial tree algorithm involving eight processes, P_0–P_7. The data to be distributed, D_0, is owned initially by P_0. The message length is the same in all steps, and the broadcast here requires three steps; the numbered arrows represent data sent in steps 1, 2, and 3, respectively. In general, the binomial tree broadcast algorithm requires $\log_2 p$ steps, where p is the number of processes.

transferred between a pair of processes. Expressed in terms of the latency α and the inverse bandwidth β, this yields the total communication cost

$$t_{\text{comm}}^{\text{one-to-all broadcast}} = \log_2 p(\alpha + l\beta) \qquad \text{[binomial tree algorithm]}. \qquad (3.1)$$

An alternative broadcast algorithm, performing the broadcast as a scatter operation followed by an all-to-all broadcast, can be modeled as follows[2]

$$t_{\text{comm}}^{\text{one-to-all broadcast}} = (\log_2 p + p - 1)\alpha + 2\frac{p-1}{p} l\beta \quad \text{[van de Geijn algorithm]}. \qquad (3.2)$$

The initial scatter, in which the process holding all the data scatters the data among all processes, requires $t_{\text{comm}} = \log_2 p\alpha + (p-1)l\beta/p$, and the following all-to-all broadcast, in which each process sends its data (of length l/p) to every other process, requires $t_{\text{comm}} = (p-1)(\alpha + l\beta/p)$ when using a ring algorithm (see Eq. 3.3). The van de Geijn algorithm has a higher latency but a smaller bandwidth term (for $p > 2$) than the binomial tree algorithm and therefore is likely to achieve better performance when broadcasting long messages for which the latency term can be neglected.

3.2.2 All-to-All Broadcast

In an all-to-all broadcast every process sends a message to every other process. The data distribution before and after an all-to-all broadcast is illustrated in Figure 3.3. This type of operation may be accomplished with a simple systolic loop algorithm (also called a ring algorithm) in which data is sent around in a virtual ring as illustrated in Figure 3.5. In each step, data exchange takes place only between neighboring processes, and every process sends data to its neighbor on one side and receives data from its neighbor on the other side.

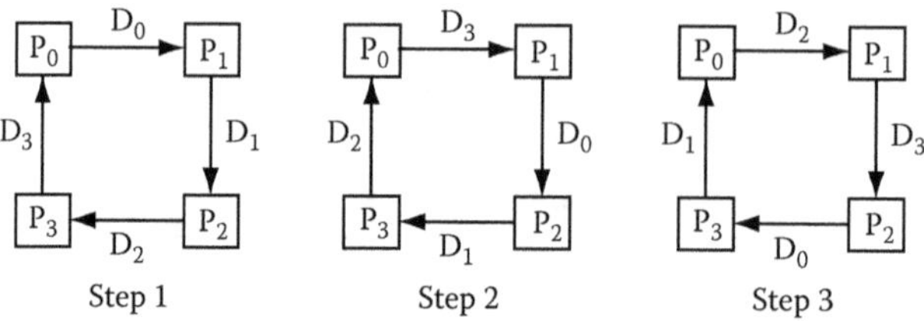

FIGURE 3.5
Data movement in systolic loop algorithm for all-to-all broadcast involving four processes, P_0–P_3. Using p processes, the algorithm requires $p - 1$ steps. Initially, process P_i owns data D_i, and upon completion, the full data set D_0, D_1, D_2, D_3 has been copied to every process.

In the first step, every process sends its own data along to the next process, and in each of the following steps every process transmits the data it received in the previous step. Using p processes, the ring algorithm requires $p-1$ steps, and the amount of data transmitted by a process in each step is constant. If l denotes the amount of data that initially resides on each process and will be transmitted to the other processes, the communication time for the all-to-all broadcast using the ring algorithm can be expressed as

$$t_{\text{comm}}^{\text{all-to-all broadcast}} = (p-1)(\alpha + l\beta) \qquad \text{[ring algorithm]}. \qquad (3.3)$$

An all-to-all broadcast can also be performed with a recursive doubling algorithm. The name of this algorithm derives from the fact that the distance between two processes exchanging information, as well as the message length, is doubled in each step of the algorithm as illustrated in Figure 3.6. In the first step, processes P_i and P_{i+1} exchange data (for even i); in the second step, P_i and P_{i+2} exchange data (for even $\lfloor i/2 \rfloor$); and in step m, process P_i exchanges data with process P_{i+2^m} (for even $\lfloor i/2^m \rfloor$). In each step, every process sends its original data as well as the data it has received from other processes in previous steps. The amount of data to be exchanged between processes thus grows with each step, and, using the data length l from above, the data to be transmitted by a process in step k can be expressed as $l2^{k-1}$. If the number of processes is a power of two, the recursive doubling all-to-all broadcast can be completed in $\log_2 p$ steps, and the required communication time is then given as

$$\begin{aligned} t_{\text{comm}}^{\text{all-to-all broadcast}} &= \log_2 p\alpha + \sum_{k=1}^{\log_2 p} 2^{k-1} l\beta \\ &= \log_2 p\alpha + (p-1)l\beta \quad \text{[recursive doubling algorithm]}. \end{aligned} \qquad (3.4)$$

If the number of processes is not a power of two, more than $\log_2 p$ steps are required in the recursive doubling algorithm, and in this case[2] the algorithm can be implemented so that the total number of steps required is bounded by $2\log_2 p$.

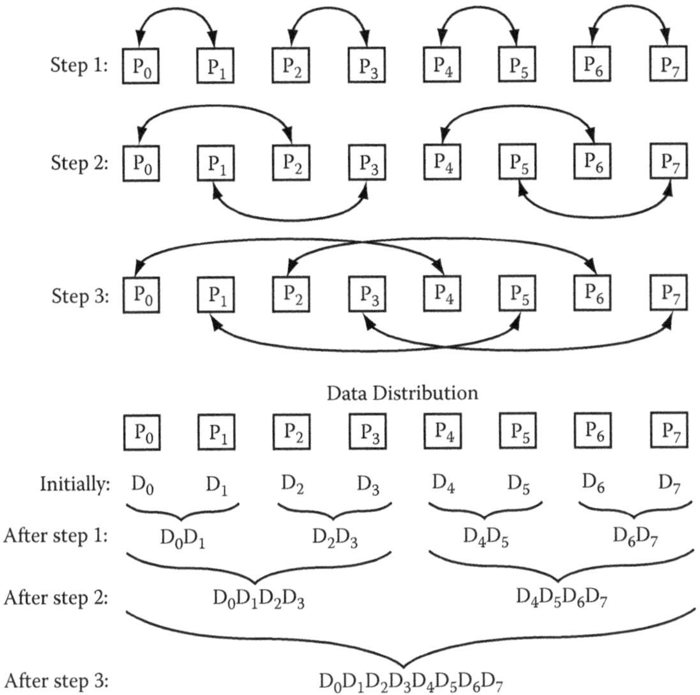

FIGURE 3.6
Communication pattern and data distribution in all-to-all broadcast using a recursive doubling algorithm involving eight processes, P_0–P_7. The top part of the figure illustrates the exchange of data between pairs of processes in each step, and the resultant data distribution in each step is shown on the bottom.

The recursive doubling algorithm can be completed in fewer steps than the ring algorithm ($\log_2 p$ vs. $p - 1$), yielding a smaller total latency, and the recursive doubling algorithm is therefore well suited for short message sizes where the latency term will tend to dominate. The bandwidth term, $(p - 1)l\beta$, is the same for the ring and the recursive doubling algorithm, and this term cannot be reduced because each process must receive an amount of data equal to $(p - 1)l$. The performance models for the broadcast given above have implicitly assumed that the connectivity of the network allows the various communication operations to be performed without causing network congestion. For this to be the case, the recursive doubling algorithm, in principle, requires a network with connectivity of at least hypercube quality (see section 2.2.4.4), although the performance model in Eq. 3.4 may be an acceptable approximation also for some networks with lower connectivity. We note that, since the bandwidth term is the same for the ring and recursive doubling algorithms, an all-to-all broadcast can be as fast on a simple ring topology as on a much more sophisticated and expensive topology if it involves long messages for which the bandwidth term dominates.

3.2.3 All-to-One Reduction and All-Reduce

The class of collective communication operations known as reduction operations involve the determination of a single result from a set of data distributed across a number of processes. Examples of reduction operations include the computation of the sum or product of a set of numbers and determination of the largest or smallest number in the set. In the following we will describe the all-to-one reduce and all-reduce operations in detail. In the case of the all-to-one reduce, the result is accumulated on one process only, whereas the all-reduce operation communicates the result of the reduction to all processes.*

The communication pattern involved in an all-to-one reduce operation is analogous to that of a one-to-all broadcast but is performed in the reverse order. Thus, whereas a message is sent from one process (the root) to all other processes in the one-to-all broadcast, all processes send a message to the root in the all-to-one reduce. In addition to message-passing, some message processing is also required locally in the all-to-one reduction. Using a binomial tree algorithm (see Eq. 3.1 and Figure 3.4), and assuming that the number of processes, p, is a power of two, the communication time can be modeled as

$$t_{\text{comm}}^{\text{all-to-one reduce}} = \log_2 p(\alpha + l\beta + l\gamma) \qquad \text{[binomial tree algorithm]}. \tag{3.5}$$

In this equation, γ represents the computational time required per word of data to perform the operation associated with the reduce (for instance, a summation or multiplication). We saw in the previous section that the one-to-all broadcast could be implemented with a bandwidth term that did not grow with the number of processes (Eq. 3.2). The trade-off for achieving the reduced bandwidth term was a larger latency term, and this approach therefore was advantageous for long messages only. Likewise, the all-to-one reduce can be performed using algorithms with β and γ terms nearly independent of the process count but with a higher latency. One such algorithm, developed by Rabenseifner,[2,3] is implemented as an all-to-all reduce followed by a gather operation, and the communication time is

$$t_{\text{comm}}^{\text{all-to-one reduce}} = 2\log_2 p\alpha + 2\frac{p-1}{p}l(\beta + \gamma/2) \qquad \text{[Rabenseifner algorithm]}. \tag{3.6}$$

For large message sizes, where the latency term can be ignored, Rabenseifner's algorithm should provide improved performance relative to the binomial tree algorithm, especially when the number of processes is large.

An all-reduce operation can be performed as an all-to-one reduction followed by a one-to-all broadcast, and, using the binomial tree algorithm for

* Note that an all-reduce operation is different from an all-to-all reduction, also called a reduce-scatter, which scatters the result of the reduction among all processes.

both the reduction and the broadcast, the communication time becomes

$$t_{\text{comm}}^{\text{all-reduce}} = \log_2 p(2\alpha + 2l\beta + l\gamma) \qquad \text{[binomial tree algorithm]}. \tag{3.7}$$

The Rabenseifner all-reduce algorithm is identical to the Rabenseifner all-to-one reduction, except for using an all-to-all broadcast (allgather) operation in place of gather, and the communication time for this algorithm can be modeled as

$$t_{\text{comm}}^{\text{all-reduce}} = 2\log_2 p\alpha + 2\frac{p-1}{p}l(\beta + \gamma/2) \qquad \text{[Rabenseifner algorithm]}. \tag{3.8}$$

Note that the performance models for Rabenseifner's all-to-one reduce and all-reduce are identical because the gather operation used for the all-to-one reduce has the same communication requirement as the all-to-all broadcast used in the all-reduce.

3.3 One-Sided Communication Operations

In *one-sided communication*, one process controls the exchange of data with another process, and data exchange does not require cooperation of the sending and receiving processes. One-sided communication can be implemented via remote direct memory access, which enables a process to access (read from or write to) the memory of another process without explicit participation from the process whose memory is accessed.

One-sided communication, like non-blocking point-to-point message-passing, is asynchronous, but one-sided communication offers the potential for improved parallel efficiency for certain types of applications. Consider, for instance, a distributed data parallel program in which processes need to access data on other processes frequently, but in a nonpredictable way. In this case, each process will be engaged in computation, and, when necessary, request data from another process. At the same time, however, every process must be able to process data requests from other processes, which may arrive at any given time. A communication scheme for this type of situation can be implemented using non-blocking send and receive operations, but some idle time will invariably result because processes actively engaged in computation are not available for taking care of incoming requests for data except at predetermined points in their computation. By using one-sided communication, however, a process can fetch the data it needs from another process without the latter process having to cooperate in the exchange, thus eliminating the wait for the serving process to respond.

Various libraries providing differing levels of support for one-sided communication operations are currently available for parallel application development. Notably, the Aggregate Remote Memory Copy Interface, ARMCI, library, which is widely used in quantum chemistry applications, provides a variety of one-sided communication operations.[4] Support for one-sided communication in MPI was specified in the MPI-2 standard.[5] The one-sided communication features in MPI-2 are subject to a somewhat complex set of rules and restrictions, and while certain implementations of MPI-2 may provide the desired functionality, other, standard-compliant, implementations might not provide acceptable performance, scalability, or functionality for quantum chemistry applications. Programmers intending to write programs that are portable to multiple MPI-2 implementations are advised to carefully study the MPI-2 specification to ensure that it meets their requirements before using the one-sided feature. A one-sided communication style can be also implemented by means of threads; we will discuss the use of multiple threads in chapter 4, and in sections 8.4 and 9.4 we will give examples of one-sided communication schemes using multiple threads.

3.4 Further Reading

Algorithms and cost analyses for a number of collective communication operations have been discussed in some detail by Grama et al.[6] A comprehensive performance comparison of implementations of MPI on different network interconnects (InfiniBand, Myrinet®, and Quadrics®), including both micro-level benchmarks (determination of latency and bandwidth) and application-level benchmarks, has been carried out by Liu et al.[7] A discussion of the optimization of collective communication in MPICH, including performance analyses of many collective operations, has been given by Thakur et al.[2]

References

1. Sur, S., H.-W. Jin, L. Chai, and D. K. Panda. RDMA read based rendezvous protocol for MPI over InfiniBand: Design alternatives and benefits. In *Proceedings of the Eleventh ACM SIGPLAN Symposium on Principles and Practice of Parallel Programming*, pp. 32–39. New York: ACM Press, 2006.
2. Thakur, R., R. Rabenseifner, and W. Gropp. Optimization of collective communication operations in MPICH. *Int. J. High Perform. C.* 19:49–66, 2005.
3. Rabenseifner, R. A new optimized MPI reduce algorithm. November 1997. http:/ / www.hlrs.de/mpi/myreduce.html.
4. The Aggregate Remote Memory Copy Interface, ARMCI, is available on the World Wide Web at http://www.emsl.pnl.gov/docs/parsoft/armci/index.html.
5. The MPI-2 standard for the Message-Passing Interface (MPI) is available on the World Wide Web at http://www.mpi-forum.org.

6. Grama, A., A. Gupta, G. Karypis, and V. Kumar. *Introduction to Parallel Computing*, 2nd edition. Harlow, England: Addison-Wesley, 2003.
7. Liu, J., B. Chandrasekaran, J. Wu, W. Jiang, S. Kini, W. Yu, D. Buntinas, P. Wyckoff, and D. K. Panda. Performance comparison of MPI implementations over InfiniBand, Myrinet and Quadrics. In *Proceedings of the 2003 ACM/IEEE conference on Supercomputing*, p. 58. Washington DC: IEEE Computer Society, 2003.

4

Multi-Threading

In the previous chapter we discussed the use of message-passing for introducing parallelism across nodes in a distributed memory parallel computer. Each node in a distributed memory computer has its own separate memory that can be accessed by the other nodes by means of message-passing. Within a given node, however, the memory is typically directly available to all of the processors on the node, and there are two principal ways to introduce parallelism: the program can execute as separate processes that each have their own independent virtual memory and share data via message-passing; alternatively, multiple threads can be used within a process. In this chapter we will discuss how to parallelize execution of a program by using multiple threads. We will outline some of the advantages and potential pitfalls of using multi-threading, compare the multi-threading and message-passing approaches, and discuss hybrid programming, which uses both multi-threading and message-passing. In Appendices B and C, the use of multi-threading will be illustrated using two different approaches, namely Pthreads and OpenMP™.

As discussed in section 2.3, a process is an abstraction within a node that provides a virtual memory separate from that of other processes. By default, a process has a single execution context, that is, a single stream of instructions is executed. Some operating systems provide support for multiple execution contexts within a single process, and these execution contexts are called *threads*. The threads can execute concurrently on different processors, or, if not enough processors are available, the operating system can serialize them onto the available processors. The relationship between processes, threads, and processors is illustrated in Figure 4.1. Threads are typically used in a *fork-join** model of computing where an initial, or main, thread is started for the process, additional threads are forked from the main thread, and, after some computation, the main thread joins with the forked threads until there is again only a single thread. The threads corresponding to a given process share resources such as heap memory (that is, memory obtained by the

* This is different from the `fork` system call in UNIX, which produces a new process providing a new execution context in a new virtual address space.

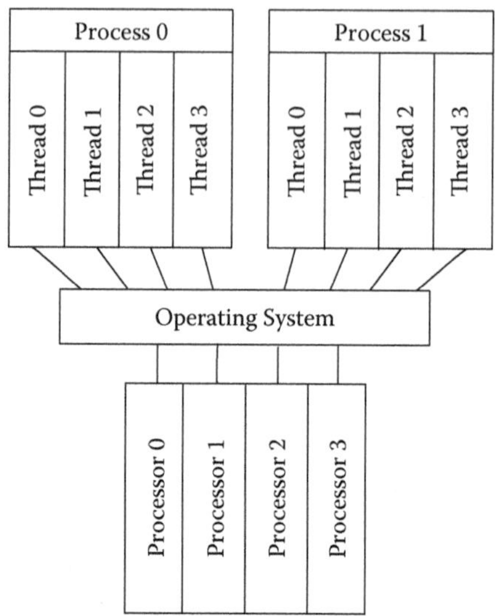

FIGURE 4.1
The relationship between processes, threads, and processors within a node. Each process has its own memory space shared among one or more threads. The threads are mapped to processors by the operating system.

C language `malloc` function or the C++ `new` operator) and file descriptors, but each thread also stores some data unique to that particular thread; for example, the address of the instruction currently being executed, the processor registers, and a stack storing temporary data generated by function calls are all specific to a given thread. It is also possible for threads to have private memory and for processes to share memory; however, for our purposes, the simplified picture presented above will suffice.

The sharing of resources between threads in the same process is both the strength and weakness of using multiple threads. Because it is trivial for multiple threads to access the shared data, data need not be communicated and replicated as it might if two processes required the same data. Thus, using multiple threads usually requires less memory than using multiple processes on a node; additionally, as soon as one thread generates new data, another thread can access that data at very low cost and without any data transfer. A disadvantage of the data sharing, however, is the need for threads to coordinate their activities whenever they modify memory to ensure that memory is always in a consistent state. We will discuss this point in more detail in sections 4.1 and 4.2.

The threads we have discussed so far are known as *kernel threads*. These threads are provided by the operating system, which can schedule different threads to run simultaneously on different processors. Another type of thread is a *user thread* (or *user-space thread*), which is scheduled by the program itself. There are two ways to perform this scheduling: an operating system timer can periodically interrupt the currently running user thread to let the program switch to another user thread; alternatively, one may require that user threads explicitly yield control, and the process can then switch execution to another user thread only at these yield points. This latter approach makes switching between threads very efficient, and the explicitly chosen yield points make it easier to ensure thread-safety. A disadvantage, however, is that a user thread with a long time period between yields can starve other threads of resources. Neither method for user threading can take advantage of multiple processors in a node. Because we are interested in using multi-threading for intra-node parallelism, our discussions of multi-threading in this book will focus on kernel threads.

4.1 Pitfalls of Multi-Threading

Because different threads in the same process share resources such as memory, extra care must be taken to ensure that these resources are manipulated in the desired way. When programming in a single-threaded environment, it can be assumed that data is fixed until it is explicitly modified; in a multi-threaded environment, however, data may change without the current thread actually modifying it—a property that can make it difficult for programmers to adjust to multi-threaded programming. Consider, for example, the program shown in Figure 4.2, which creates two threads that each compute an integer value and add that value into the same memory location, `datum`. If the two threads execute their code serially, we achieve the expected result, but if both threads are active concurrently, or their execution is interleaved on the same processor, then one of several outcomes can occur. Three of the possible outcomes, each producing a different result, are shown in Table 4.1. In Case 1, Thread

```
Integer datum = 0
Create Thread 1 which computes:
    Integer a = compute_a()
    datum = datum + a
Create Thread 2 which computes:
    Integer b = compute_b()
    datum = datum + b
Wait for threads to complete
```

FIGURE 4.2
Pseudocode illustrating a race condition in a multi-threaded program. Two threads are created, each reading and modifying a datum at the same memory location, datum.

TABLE 4.1

Outcomes for three of the possible code paths that may arise due to the race condition in Figure 4.2

Case 1		Case 2		Case 3	
Thread 1	**Thread 2**	**Thread 1**	**Thread 2**	**Thread 1**	**Thread 2**
read datum		read datum		read datum	
datum+=a			read datum		read datum
write datum		datum+=a		datum+=a	
	read datum		datum+=b		datum+=b
	datum+=b	write datum			write datum
	write datum		write datum	write datum	
outcome: datum = a+b		outcome: datum = b		outcome: datum = a	

1 finishes execution before Thread 2 starts, and by the time Thread 2 reads `datum`, Thread 1 has completed its update to `datum`, and the expected result is obtained. In Case 2, both threads first read a copy of `datum`, perhaps loading its value into a register. Each thread then adds its result to this temporary copy and writes out the result, with Thread 2 writing out `datum` last. In this case, Thread 2 overwrites the contribution previously written out by Thread 1, and the result is incorrect. Case 3 is similar to Case 2, but when the datum is written out by the two threads, the order of the writes is reversed, and the `datum` written out lacks the contribution computed by Thread 2. The situation illustrated in Table 4.1 is called a *race condition*, and special precautions must be taken to avoid such cases.

Let us briefly consider how one can avoid a race condition in a program like the one shown in Figure 4.2. The race condition is caused by the need for each thread to create a temporary copy of the datum to be able to add a contribution to it. If the temporary copy were in a processor cache supporting cache-coherency, this would not present a problem because the node's cache-coherency protocol would ensure consistency across all threads. In our example, however, there is a period of time while the datum is being processed during which the cache-coherency mechanism cannot prevent simultaneous use of the datum by another thread because the datum is kept in a register. Regions of a program that can be safely executed by only one thread at a time are known as *critical sections*. The programmer must specifically identify critical sections and ensure that they are correctly handled. One way to handle a critical section is to acquire a *mutual exclusion lock*, or *mutex*, that permits only one thread to execute a given region of code at a time. Exactly how this is done depends on the particular multi-threading implementation used, and it will be illustrated for Pthreads in Appendix B and for OpenMP in Appendix C.

What would be the consequences of overlooking the critical section where `datum` is accumulated in the example from Figure 4.2? If `compute_a` were a function that usually ran faster than `compute_b`, the outcome, in most cases, would be that of Case 1 of Table 4.1, which is the correct answer. Thus, a problem of this type could go undetected, even though the program is

```
Create Mutex A
Create Mutex B
Create Thread 1, which executes:
    Lock mutex A
    Lock mutex B
    Process resources protected by A and B
    Unlock mutex B
    Unlock mutex A
Create Thread 2, which executes:
    Lock mutex B
    Lock mutex A
    Process resources protected by A and B
    Unlock mutex A
    Unlock mutex B
Wait for threads to complete
```

FIGURE 4.3
An example of a flawed locking protocol that can result in deadlock.

non-deterministic and sometimes will produce the wrong answer. Consequently, it can be very difficult to test and debug multi-threaded code. A combination of good program design and tools for detecting unprotected critical sections are essential components of multi-threaded programming.

Another hazard that must be dealt with when programming for multiple threads is the possibility of *deadlocks*. Deadlocks are situations where one or more threads try to acquire a lock on an already locked mutex that will never be unlocked due to an error in the programmer's use of mutexes. An example of a potential deadlock is shown in Figure 4.3. Here, two mutexes are created, A and B, each protecting some shared resources that are unspecified in this example. Two threads are then created, Thread 1 and Thread 2, both requiring access to the resources protected by mutexes A and B. Thread 1 first acquires a lock on mutex A and then B. Thread 2 reverses this order, first locking mutex B and then A. Unfortunately, it is possible for Thread 1 to lock A and Thread 2 to lock B before Thread 1 is able to lock B. In this case, neither Thread 1 nor Thread 2 can proceed, and the program will deadlock. A problem like this can be avoided by ensuring that all threads that lock multiple mutexes use a consistent locking order.

Deadlock can also result from using a mutex in a recursive routine. Consider, for instance, a routine that acquires a lock on a mutex and then calls itself without unlocking the mutex. If the routine then attempts to lock the mutex again, the thread will typically deadlock because the mutex is already locked. This will not happen, however, if a special mutex explicitly supporting recursion is used; in this case, a given thread can hold more than one lock on a given mutex. Note that recursive mutexes usually have more overhead than the nonrecursive variety.

4.2 Thread-Safety

The concept of *thread-safety* refers to the use of various programming techniques for creating code that works properly in a multi-threaded environment. A thread-safe function will act correctly when it is executed concurrently by multiple threads. Related is the concept of a *reentrant function*, which depends only on its arguments and uses only temporary values allocated on the stack as well as the results of calls to other reentrant functions. Reentrant functions are also thread-safe (although it is possible to use a reentrant function in an unsafe way if it modifies data passed by reference, and the programmer passes the same data reference to the reentrant function from multiple simultaneously executing threads). Many of the C library calls keep data in global memory, and, thus, are not reentrant. Neither are they thread-safe because they do not obtain a mutual exclusion lock to protect their global data. The `drand48` pseudo-random number generator is an example of a nonreentrant, nonthread-safe function, because it produces its result by reading and modifying global data without obtaining a mutual exclusion lock. The C library provides reentrant versions of such routines. They are identified by the `_r` suffix in their function name and have a different signature requiring that all needed data is passed as an argument. In the `drand48` case, the `drand48_r` function takes two arguments: a pointer to data of type `struct drand48_data`, which contains the needed state information, and a pointer to `double` in which the result is placed. Each thread calling `drand48_r` must provide its own storage for each of these arguments to use the function in a thread-safe fashion.

Many of the techniques for writing thread-safe code are well aligned with what is generally considered good programming practice. In the discussion of the pitfalls of multi-threaded programming in the previous section we briefly discussed how to avoid race conditions and deadlocks, and below we list a few general guidelines to help ensure the thread-safety of code:

- Make functions reentrant when possible.
- Use caution when sharing data that could be modified:
  - Avoid global and static data.
  - When possible, avoid modifying data to which multiple threads have pointers.
- Use mutual-exclusion locks:
  - When modifying shared data.
  - When calling functions that are not thread-safe.

- Care must be taken to avoid deadlocks:
  - A recursive function should only acquire a mutex once, or a mutex supporting recursion must be used.
  - Pay attention to locking/unlocking order when using multiple mutexes.

4.3 Comparison of Multi-Threading and Message-Passing

Let us compare the message-passing and multi-threading approaches by considering a parallel algorithm for performing a matrix–vector multiplication, $\mathbf{c} = \mathbf{Ab}$. In this example, $\mathbf{A}$ is an $n \times n$ matrix, the number of processes is designated n_{proc}, and we will assume that n is a multiple of n_{proc}. The message-passing version of this algorithm is implemented using the Message-Passing Interface (MPI) and shown in Figure 4.4.[†] Parallelism is achieved by using concurrently executing processes, each with their own address space. Each process has a copy of the entire $\mathbf{b}$ vector, while the $\mathbf{A}$ matrix is distributed by rows among all the processes; process i holds the n/n_{proc} rows numbered from $i \times n/n_{\text{proc}}$ to $(i+1)n/n_{\text{proc}} - 1$, and the portions specific to each process are stored in $\mathbf{A}_{\text{local}}$. Every process uses the locally stored part of $\mathbf{A}$ to compute a portion of the $\mathbf{c}$ vector, $\mathbf{c}_{\text{local}} = \mathbf{A}_{\text{local}}\mathbf{b}$. To construct the complete $\mathbf{c}$ vector, an `MPI_Allgatherv` call is employed to broadcast each process's $\mathbf{c}_{\text{local}}$ into a complete $\mathbf{c}$ vector that is the same on all processes.

A multi-threaded analogue of this program, shown in Figure 4.5, is implemented using Pthreads.[‡] When the `mxv` routine is entered, there is only a single thread of control, the master thread. This thread sets up a small amount of context information, `work`, that will be used by each worker thread and contains references to the data, the size of the problem, and the work that each thread is to execute. Threads are started by calling `pthread_create`, passing in as arguments the `work` context and a routine, `run_thread`, that uses the context to perform work. Before proceeding, the master thread awaits the completion of the work by calling `pthread_join` for each thread. At this point in the program, the $\mathbf{c}$ vector is complete, and the computation can proceed. No communication is required in this algorithm because the threads share a single address space, making all the data immediately available to all threads.

Another multi-threaded implementation of the matrix–vector multiplication, using OpenMP,[§] is shown in Figure 4.6. This implementation is much

---

[†] Message-passing and MPI are discussed in chapter 3 and Appendix A, respectively. Sections 5.3.2 and 6.4.1 present performance analyses of this algorithm.

[‡] Pthreads is discussed in Appendix B.

[§] OpenMP is discussed in Appendix C.

```
void mxv(double *c, double **A_local, double *b, int n,
         int local_n) {
  double *c_local;
  int *n_on_proc = (int*)malloc(n*sizeof(int));
  int *offset_on_proc = (int*)malloc(n*sizeof(int));
  int i, nproc;
  MPI_Comm_size(MPI_COMM_WORLD,&nproc);
  for (i=0; i<nproc; i++) {
      n_on_proc[i] = local_n;
      offset_on_proc[i] = i*local_n;
  }
  c_local = (double*)malloc(sizeof(double)*local_n);
  for (i=0; i<local_n; i++)
      c_local[i] = dot(A_local[i],b,n);
  MPI_Allgatherv (c_local, local_n, MPI_DOUBLE,
                  c, n_on_proc, offset_on_proc,
                  MPI_DOUBLE, MPI_COMM_WORLD);
  free(c_local);
}
```

FIGURE 4.4
A routine to compute the matrix vector product $\mathbf{c} = \mathbf{Ab}$ using message-passing. The input data `b` is replicated, and `A` is distributed with local rows in `A_local`. The output data `c` is replicated among all of the processes. The number of processes is designated `nproc`, and `local_n` is the number of elements of the `c` vector that will be computed locally.

simpler than the Pthreads version from Figure 4.5, and the OpenMP version differs from the scalar code only by the addition of the OpenMP pragma.

4.4 Hybrid Programming

While multi-threading provides parallelism within a single shared memory node, algorithms running on multiple nodes of large-scale parallel computers must use message-passing to explicitly move data between nodes. A hybrid programming technique can be used to utilize multi-threading for single-node parallelism in combination with message-passing for parallelization across many nodes. This hybrid approach entails all the complexity of both multi-threading and message-passing as well as some additional concerns that will be outlined next. The advantages of the hybrid approach relative to pure message-passing is the fast synchronization within a node, the data sharing between threads, and an overall smaller memory requirement. These features permit a more fine-grained intra-node parallelism than what could be used with separate processes in a message-passing scheme and will make it easier, or possible, to utilize extremely large computers.

```
typedef struct {
    double *c, *b, **A;
    int begin, end, n;
} work_t;
void *run_thread(void *arg) {
  work_t *work = (work_t*)arg;
  int i;
  for (i=work->begin; i<work->end; i++)
      work->c[i] = dot(work->A[i],work->b,work->n);
  return 0;
}
void mxv(double *c, double **A, double *b, int n,
          int nthread) {
  int i;
  work_t *work = malloc(sizeof(work_t)*nthread);
  pthread_t *thread = malloc(sizeof(pthread_t)*nthread);
  for (i=0; i<nthread; i++) {
    work[i].c = c;
    work[i].b = b;
    work[i].A = A;
    work[i].n = n;
    work[i].begin = (n/nthread)*i;
    work[i].end = work[i].begin + n/nthread;
    pthread_create(&thread[i],NULL,run_thread,&work[i]);
    }
  for (i=0; i<nthread; i++) {
      pthread_join(thread[i],0);
    }
  free(work);
  free(thread);
}
```

FIGURE 4.5
A routine to compute the matrix vector product $\mathbf{c} = \mathbf{Ab}$ using Pthreads.

```
void mxv(double *c, double **A, double *b, int n) {
  int i;
#pragma omp parallel for
  for (i=0; i<n; i++)
      c[i] = dot(A[i],b,n);
}
```

FIGURE 4.6
A routine to compute the matrix vector product $\mathbf{c} = \mathbf{Ab}$ using OpenMP.

TABLE 4.2

The four levels of multi-threading support defined by the MPI standard

Single	The application may use only a single thread.
Funneled	All MPI calls must be made from the initial thread of a process.
Serialized	Only a single thread may call an MPI routine at a time. The programmer must ensure that this condition is met.
Multiple	Fully thread-safe. Multiple threads may concurrently be involved in MPI calls.

As discussed earlier in this chapter, thread-safety of functions is a key issue when using multi-threading, and in hybrid programming models, thread-safety of message-passing libraries is an important concern as well. The Message-Passing Interface (MPI) defines four different levels of support for multi-threading, shown in Table 4.2. The programmer can specify the desired level of multi-threading support by using `MPI_Init_thread` to initialize MPI. This routine is given a flag that specifies the desired level of multi-threading support and an output variable in which will be placed the actual level of multi-threading support that MPI will provide. If the `MPI_Init` routine is used to initialize MPI, then the level of threading support is implementation defined. MPI implementations are not required to support all of the multi-threading levels, and those that are supported by a given MPI implementation may not perform well. For example, some MPI implementations that utilize hardware support for remote direct memory access detect incoming messages by polling; polling entails repeatedly reading a specific memory location until its value changes, and using this technique for detecting incoming messages yields very high performance for micro-benchmarks. However, if we write an application that requires a thread to repeatedly use an `MPI_Read` call to wait for infrequently arriving data, then, if a polling MPI implementation is used, the thread in question will be continuously occupied by polling; hence, the processor, on which this thread is running, will not otherwise be available for computation. This could lead to a significant performance loss, although in a node using manycore technology, the effective loss of one processor for computation might be acceptable. Alternatively, an MPI implementation that employs interrupts to signal the arrival of new messages could be used.

In Figure 4.7 we show a hybrid matrix–vector multiplication algorithm obtained by combining the MPI parallelization shown in Figure 4.4 with the OpenMP parallelization shown in Figure 4.6. There are two levels of parallelism: using MPI, `nproc` processes are first started up, one on each node, and each of these processes then spawns a number of threads by means of the OpenMP pragma. This implementation required addition of just a single line of code to the MPI version, and the resulting program will work with any MPI that provides the "funneled" level of multi-threading support.

Figure 4.8 shows a more complex hybrid programming example. It depicts the threads in the second-order Møller-Plesset perturbation theory algorithm P2 discussed in section 9.4. This algorithm sets up both computation threads and communication threads on each node. The computation threads perform

```
void mxv(double *c, double **A_local, double *b, int n,
         int local_n) {
  double *c_local;
  int *n_on_proc = (int*)malloc(n*sizeof(int));
  int *offset_on_proc = (int*)malloc(n*sizeof(int));
  int i, nproc;
  MPI_Comm_size(MPI_COMM_WORLD,&nproc);
  for (i=0; i<nproc; i++) {
      n_on_proc[i] = local_n;
      offset_on_proc[i] = i*local_n;
  }
  c_local = (double*)malloc(sizeof(double)*local_n);
#pragma omp parallel for
  for (i=0; i<local_n; i++)
      c_local[i] = dot(A_local[i],b,n);
  MPI_Allgatherv (c_local, local_n, MPI_DOUBLE,
                  c, n_on_proc, offset_on_proc,
                  MPI_DOUBLE, MPI_COMM_WORLD);
  free(c_local);
}
```

FIGURE 4.7
A routine to compute the matrix vector product $\mathbf{c} = \mathbf{Ab}$ using a hybrid multi-threading and message-passing technique. The algorithm shown combines the MPI and OpenMP parallelizations illustrated in Figures 4.4 and 4.6. Since all MPI calls are from the main thread, MPI need only support the "funneled" level of multi-threading.

the bulk of the floating point intensive work, computing the four-index atomic orbital integrals and transforming them from the atomic to the molecular orbital basis. After two quarter transformations are performed, a redistribution of the data is required. The redistribution is accomplished by the compute thread's send call, and the recipient of this send is a communication thread on a remote node, which receives the data and sums the contribution into the appropriate memory location. This algorithm requires the "multiple" level of multi-threading support in MPI because multiple computation threads may be sending data simultaneously and also because the communication threads can receive data while the computation threads are sending and other communication threads are receiving.

4.5 Further Reading

Appendix B briefly discusses programming with Pthreads, and the text by Lewis and Berg[1] is a good source for more information on multi-threaded programming with Pthreads. Appendix C gives a brief introduction to

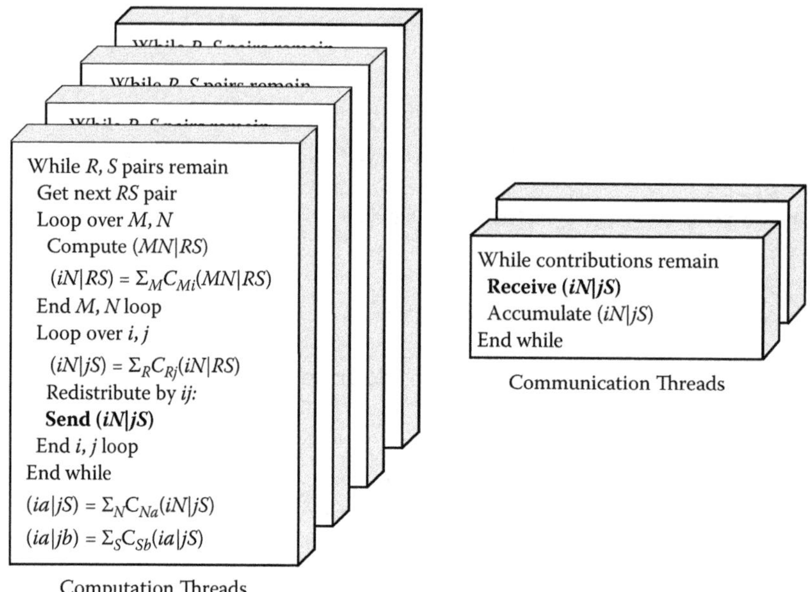

FIGURE 4.8
Computation and communication threads in a parallel MP2 algorithm (the algorithm shown is the P2 algorithm discussed in section 9.4). Steps involving communication are shown in boldface.

OpenMP, and the OpenMP standard[2] includes complete details for programming with OpenMP.

References

1. Lewis, B., and D. J. Berg. *Multithreaded Programming with PThreads*. Upper Saddle River, NJ: Prentice Hall, 1997.
2. OpenMP Application Program Interface. Version 2.5, May 2005. http://www.openmp.org.

5

Parallel Performance Evaluation

While the performance of a scalar algorithm is usually measured in terms of execution time and resource requirements, such as memory and disk usage, additional performance metrics are needed for parallel algorithms. Several factors, including communication overhead, degree of parallelism, and load imbalance, must be incorporated into parallel performance measures to be able to characterize the performance of a parallel algorithm over a wide range of parallel computers and process* counts. The development of a performance model capable of making realistic predictions of the parallel performance and exposing potential shortcomings of the algorithms should be an integral part of parallel program development.

In this chapter we will consider issues pertaining to parallel performance modeling. We first introduce some network performance characteristics for parallel computers that must be considered when modeling parallel performance. We then present several performance measures for parallel programs, and we discuss how to develop a performance model for a parallel algorithm. Finally, we will discuss how to evaluate performance data and illustrate how reported performance data can be potentially misleading.

5.1 Network Performance Characteristics

The network performance characteristics for a parallel computer may greatly influence the performance that can be obtained with a parallel application. The *latency* and *bandwidth* are among the most important performance characteristics because their values determine the communication overhead for a parallel program. Let us consider how to determine these parameters and how to use them in performance modeling. To model the communication time required for a parallel program, one first needs a model for the time required to send a message between two processes. For most purposes, this time can

*We use the term "process" to refer to an operating system process along with one or more dedicated processors.

be modeled using an *idealized machine model*, which assumes that the time depends on the length of the message but is independent of both the relative locations of the two processes and other concurrent network traffic. Using the idealized machine model, the total time required to send a message between two processes can be expressed as follows

$$t_{\text{send}} = \alpha + l\beta \tag{5.1}$$

where α represents the latency, β is the inverse bandwidth, and l denotes the message length. When a process posts a sending operation to send a message to another process, a certain amount of time is required to prepare the message to be sent and to propagate the first byte of the message through the network; this lag is the latency, or startup time, and it represents the time that elapses before any data is received.[†] Using the idealized machine model, the time required to complete the transmission, in addition to the startup time, is proportional to the message length. The proportionality constant is β, the inverse of the bandwidth. The bandwidth, β^{-1}, is the transfer rate and is typically measured in Mbytes/s.

Another network performance characteristic, related to the latency and bandwidth, is the *effective bandwidth*, which is defined as the message length divided by the total send time

$$\text{Effective bandwidth} = \frac{l}{t_{\text{send}}}. \tag{5.2}$$

The effective bandwidth is a measure for the overall rate of transfer for a message, including the startup time, and it increases with the message size until reaching an asymptotic value, which equals the bandwidth β^{-1}. In Figure 5.1, the time required to send a message between two processes is plotted as a function of the message size; the resulting effective bandwidth is shown as well. The values for α and β can easily be obtained from the plots: as the message size approaches zero, the communication time approaches α; as the message size grows large, the effective bandwidth asymptotically approaches β^{-1}. In the figure, we also show the communication time predicted by the idealized machine model (Eq. 5.1), demonstrating that this model accurately predicts the measured communication time over the entire range of message sizes included. The values of α and β can also readily be obtained from a linear plot of the communication time versus the message size: this plot will be a straight line whose slope and intercept with the y-axis represent β and α, respectively.

When evaluating parallel network performance, it is important to ascertain whether reported bandwidth data refer to unidirectional or bidirectional bandwidths. Most communication networks provide bidirectional communication channels, which are able to transmit messages in both directions

[†] Note that the term "latency" is sometimes used differently in the parallel computing literature; in some texts the latency represents the time required to send a message (i.e., t_{send}).

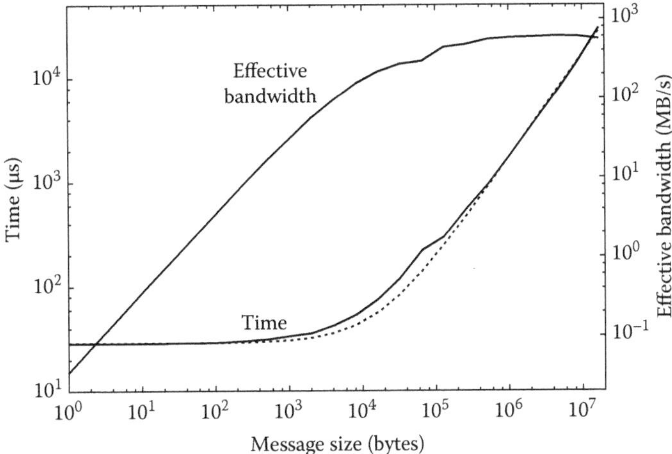

FIGURE 5.1
Log–log plot of the time required to send a message between two processes as a function of message size; the resulting effective bandwidth is shown as well. The dashed line represents the linear function $\alpha + l \times \beta$, where l is the message size in bytes, and the values of α and β are 29 μs and (590 MB/s)$^{-1}$, respectively. Data were obtained with the program shown in Figure 5.2 running on a Linux cluster[12] using an InfiniBand interconnect employing IPoIB.

simultaneously. A bandwidth measured by sending data in only one direction at a time is referred to as a *unidirectional* bandwidth, whereas a *bidirectional* bandwidth is obtained by measuring transfer rates while simultaneously sending data in both directions on a communication channel. When bidirectional bandwidths are reported, the total amount of data going in both directions is counted, so bidirectional bandwidths are usually nearly twice as large as unidirectional bandwidths (in practice, there is normally a performance drop when data is transmitted in both directions at the same time). Performance models based on the send time for a single message given in Eq. 5.1 are formulated in terms of an inverse bandwidth, β, that pertains to the rate of data transfer in one direction; these models, therefore, should use a β value based on the unidirectional bandwidth, or, alternatively, one-half times the bidirectional bandwidth.

The latency and bandwidth are machine-specific parameters, and they depend on both the message-passing hardware and software employed. Examples of values for these parameters, which may differ widely for different communication networks, are listed in Table 5.1. As mentioned (in Figure 5.1), the latency and bandwidth can be determined for a given parallel computer by measuring the time required for sending messages between processes as a function of the message size. In Figure 5.2 we show a program that can be used for this purpose, and this program was used to generate the data presented in Figure 5.1. Programs for determining α and β are also available on the World Wide Web.[1] The program shown in Figure 5.2 employs MPI

TABLE 5.1

Latency α, inverse bandwidth β, and bandwidth β^{-1} for Gigabit Ethernet (GigE) and InfiniBand (using IPoIB) interconnects on a Linux cluster.[12] Data were determined using the program shown in Figure 5.2, and the reported bandwidths are unidirectional

	α (μs)	β (ns/byte)	β^{-1} (Mbytes/s)
GigE	143	8.5	118
IPoIB	29	1.7	590

blocking point-to-point send and receive operations, and it can be used on any parallel computer on which MPI is available. The program calls the function `pingpong`, which sends data back and forth between two processes, and measures the corresponding communication time as a function of the message size. To ensure that the measured communication times are sufficiently long to be meaningful (that is, much greater than the resolution of the system clock), timings are recorded for a large number of repeated calls of the `pingpong` function. The times thus measured are divided by the number of repetitions and by a factor of two to obtain the time required to send a message between the two processes. Timings are printed out for each message size, and the measured values of α and β are printed out as well. Note that this program measures the unidirectional bandwidth: although data is sent back and forth between two processes, data transfer does not take place in both directions at the same time.

5.2 Performance Measures for Parallel Programs

5.2.1 Speedup and Efficiency

One of the most widely used performance measures for parallel programs is the *speedup*, $S(p)$, which is defined as

$$S(p) = \frac{t(1)}{t(p)} \tag{5.3}$$

where $t(1)$ and $t(p)$ denote the execution time when running on a single process and on p processes, respectively. The *ideal speedup*, also referred to as *linear speedup*, generally equals the number of processes employed, $S(p) = p$. Another performance measure, closely related to the speedup, is the efficiency, $E(p)$, defined as

$$E(p) = \frac{S(p)}{p}. \tag{5.4}$$

```
#include <mpi.h>
#include <stdio.h>

static inline void pingpong(int me, char *a, int na) {
  MPI_Status status;
  if (me == 0) {
    MPI_Send(a, na, MPI_CHAR, 1, 0, MPI_COMM_WORLD);
    MPI_Recv(a, na, MPI_CHAR, 1, 0, MPI_COMM_WORLD, &status);
  }
  else if (me == 1) {
    MPI_Recv(a, na, MPI_CHAR, 0, 0, MPI_COMM_WORLD, &status);
    MPI_Send(a, na, MPI_CHAR, 0, 0, MPI_COMM_WORLD);
  }
}
int main(int argc, char **argv) {
  const int maxdata = 1<<24;
  int i, j, me;
  double t, alpha, beta;
  void *a = (void*)malloc(maxdata);
  int nrepeat=100000;
  memset(a,0,maxdata);
  MPI_Init(&argc, &argv);
  MPI_Comm_rank(MPI_COMM_WORLD, &me);
  for (i=0; i<=maxdata; i=(i?i*2:1)) {
    double t1, t0;
    pingpong(me,a,i);
    t0 = MPI_Wtime();
    for (j=0; j<nrepeat; j++) pingpong(me,a,i);
    t1 = MPI_Wtime();
    if (me == 0) {
      t = 0.5*(t1-t0)/nrepeat;
      printf("%10d %15.9f %15.3f\n", i, t, i/t);
      if (i==0) alpha = t;
      if (i==maxdata) beta = (t-alpha)/i;
    }
    if (i>1000 && nrepeat>10) nrepeat /= 2;
  }
  if (me == 0)
    printf("alpha = %12.9f sec\n1/beta = %12.3f bytes/sec\n",
           alpha, 1/beta);
  MPI_Finalize();
  return 0;
}
```

FIGURE 5.2
A program to estimate latency, α, and inverse bandwidth (unidirectional), β, for a parallel computer. The program is written in the C programming language and employs MPI blocking send and receive operations. For each message size `i`, the program calls the function `pingpong`, which sends a message back and forth between two processes, and measures the time required for this message transfer. The `pingpong` function is called `nrepeat` times for each message size to ensure that the measured time is much greater than the resolution of the system clock.

The efficiency measures how well the computer is utilized by a parallel application, and the efficiency is usually expressed as a percentage, with an ideal efficiency corresponding to $E(p) = 100\%$.

In general, the highest speedup that can by obtained by a parallel algorithm is $S(p) = p$. This upper bound on the speedup applies when the algorithm requires the same total number of operations when using a single process and p processes, and the computers used for the single-process and p-process timings match in all aggregate properties. Thus, the total amount of memory and disk space available as well as the bandwidth for accessing these resources should be the same for the single-process and the p-process computations. If these conditions are not met, it is possible to get *superlinear* speedups, $S(p) > p$. In quantum chemistry, for instance, integral-direct algorithms are an important class of algorithms for which superlinear speedups are commonly encountered. These algorithms will store a number of the integrals (depending on the available memory) and recompute the rest of the integrals as they are needed; when running on a distributed memory computer where the aggregate memory increases with the number of processes, these algorithms can take advantage of the increase in available memory to reduce the number of integrals that must be recomputed, effectively reducing the total work and making superlinear speedups possible. We will illustrate examples of such superlinear speedups in section 5.4. Whereas integral direct algorithms by design take advantage of the increased memory to achieve superlinear speedups, other classes of algorithms may produce superlinear speedups as well. For example, algorithms that use distributed data will be able to fit a larger fraction of the data in cache as the number of processes increases; the total amount of data that must be fetched from slower memory will therefore decrease, possibly leading to superlinear speedups.

When considering the speedup for a given parallel algorithm, $t(p)$ (with $p > 1$) always represents the execution time for that particular algorithm, although $t(1)$ may sometimes represent a timing for a different algorithm. Thus, $t(1)$ can be the single-process execution time for the fastest existing scalar algorithm, and the resulting speedup is then designated the *absolute* speedup. Alternatively, $t(1)$ may represent the execution time of the current parallel algorithm when running on one process. In this case, the computed speedup will be a *relative* speedup, which measures how well the algorithm has been parallelized but not does provide any information about the absolute performance gain achieved by the parallelization. The efficiency, like the speedup, may be relative or absolute depending on the definition of $t(1)$. Throughout this book, we will use the terms speedup and efficiency without specifying whether the quantity in question is relative or absolute, unless this distinction is important. In computational science applications, the speedups and efficiencies considered are often relative. Absolute speedups and efficiencies are not easily obtained for quantum chemistry algorithms, for instance, because it may be difficult to determine which algorithm is the best existing scalar algorithm. The fastest scalar algorithm may not be the same for different test cases or for different computers, or it may perhaps be part of a commercial package that is not available to the programmer. Additionally, speedup curves are usually employed to assess how well an algorithm has been parallelized, and for this purpose relative speedups tend to be more informative than absolute speedups.

In practice, ideal speedups are difficult to achieve, especially if the number of processes is large. One factor that reduces the speedup is the existence in an algorithm of inherently sequential parts of code that cannot benefit from a parallel implementation. An upper bound on the speedup was formulated by Amdahl,[2] who expressed the maximum attainable speedup for a parallel algorithm in terms of the serial fraction, f, of the algorithm

$$S(p) \leq \frac{1}{f + (1 - f)/p} \leq \frac{1}{f}. \tag{5.5}$$

In this relation, known as *Amdahl's law*, f represents the fraction of the single-process execution time that is consumed by the parts of the algorithm that have not been parallelized, and f is defined as

$$f = \frac{t_s}{t_s + t_p} \tag{5.6}$$

where t_s and t_p denote the single-process execution time for the serial parts and the parallel parts of the program, respectively. It follows from Amdahl's law that even a small fraction of serial code can seriously reduce the maximum obtainable speedup: for example, if $f = 1/10$, the speedup cannot exceed 10 no matter how many processes are employed.

The speedup limit of $1/f$ given by Amdahl's law is derived using the assumption that the fraction of serial code is independent of the problem size. It has been argued by Gustafson,[3] however, that the serial fraction is likely to decrease when the problem size increases, and that the definition of the speedup should reflect this. Gustafson expressed the execution time on p processes and on a single process as follows

$$t_G(p) = t_s + t_p(p) \tag{5.7}$$

$$t_G(1) = t_s + p \times t_p(p) \tag{5.8}$$

where $t_p(p)$ designates the execution time on p processes of the parallel part of the algorithm. This leads to the following so-called *scaled speedup*, here designated S_G,

$$S_G(p) = \frac{t_G(1)}{t_G(p)} = \frac{t_s + p \times t_p(p)}{t_s + t_p(p)}. \tag{5.9}$$

Note that the t_p used by Gustafson, as defined by Eq. 5.7, is different from the t_p employed in Eq. 5.6. Defining the fraction of serial code in terms of the execution time on p processes

$$f_G = \frac{t_s}{t_s + t_p(p)} \tag{5.10}$$

the following expression, known as *Gustafson's law*, gives the resulting speedup

$$S_G(p) = f_G + p \times (1 - f_G) = p + (1 - p) \times f_G. \tag{5.11}$$

By using a definition of f_G that depends on p, Gustafson's law predicts a speedup without a fixed upper bound. Note, however, that the speedup $S_G(p)$ is not a linear function of p, as it may appear from Eq. 5.11, because f_G also depends on p.

It is important to realize that Amdahl's and Gustafson's laws are not contradictory and that different expressions for the speedup are obtained simply because different definitions are employed for the serial fraction of code, f and f_G, respectively. We have included a discussion of Amdahl's and Gustafson's laws because they are frequently invoked in the literature in discussions of parallel computing; these laws provide simple upper bounds to attainable parallel speedups, but they are generally of limited value for parallel performance analysis. The speedup limits given by Amdahl's and Gustafson's laws are obtained by assuming that a fraction of the code is inherently sequential, and that the remaining part of the code can be parallelized in a way that yields ideal speedups. In programs that have been parallelized from the outset, however, the sequential fraction of code (whether computed from Amdahl's or Gustafson's definition) can often be made so small that its effect on the speedup is essentially negligible. In practice, the major obstacles to achieving linear speedups are factors such as communication overhead and load imbalance, and these factors must be taken into account when modeling parallel performance.

In Figure 5.3 we show a variety of speedup curves illustrating some parallel performance characteristics often encountered in practice. The ideal speedup curve (a) is a straight line with unit slope (for a program with constant efficiency, the speedup curve will always be a straight line). The speedup curve (b) represents a case with superlinear speedups but degrading performance due to factors such as communication overhead or load imbalance. This curve initially rises above the ideal speedup curve but drops below the ideal curve as the number of processes increases. The speedup curves (c) and (d) both represent a case in which the computation time scales ideally, but the communication overhead causes degrading performance. In (c), the communication overhead is logarithmic, and relatively high speedups are obtained even for large process counts, whereas in (d), the communication overhead increases linearly with the number of processes, resulting in rapid performance degradation and a speedup curve that turns downward for larger process counts. Finally, the speedup curve (e) illustrates speedups obtained for an algorithm that has a small serial fraction (0.025) but is otherwise parallelized with no load imbalance or communication overhead.

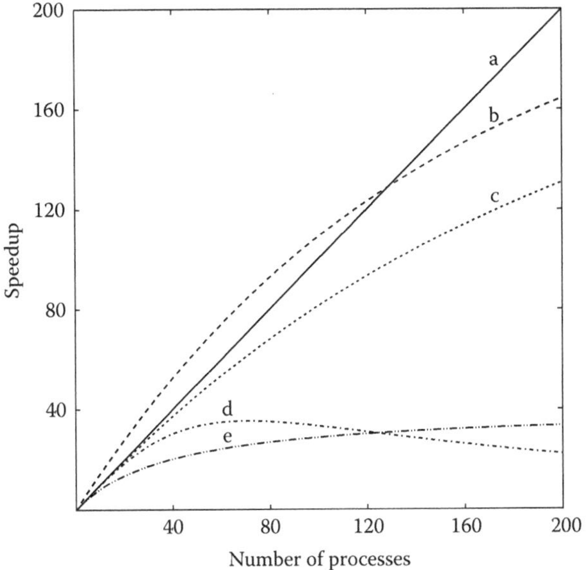

FIGURE 5.3
Speedup curves illustrating commonly encountered performance patterns. (a) ideal; (b) superlinear speedup with performance degradation due to, e.g., communication overhead or load imbalance; (c) logarithmic communication overhead; (d) linear communication overhead; (e) incompletely parallelized program (serial fraction of 0.025). See text for details.

5.2.2 Scalability

A parallel algorithm is said to be *scalable* if its parallel efficiency can be maintained as the number of processes increases. Scalability may be classified as either weak or strong, depending on how the efficiency varies with the number of processes and the problem size. An algorithm is *weakly scalable* if the efficiency does not decrease as the number of processes grows, provided that the problem size increases as well. Thus, for process counts p_1 and p_2 and problem sizes n_1 and n_2, with $p_2 > p_1$ and $n_2 > n_1$, weak scalability implies that

$$E(p_2, n_2) \gtrsim E(p_1, n_1) \tag{5.12}$$

where $E(p, n)$ is the efficiency (see Eq. 5.4), here expressed as a function of both the number of processes and the problem size. If the efficiency meets the more rigorous requirement of being roughly constant as the number of processes increases for a fixed problem size, that is,

$$E(p_2, n) \approx E(p_1, n) \tag{5.13}$$

the algorithm is said to be *strongly scalable*. Expressed in terms of speedup, strong scalability implies that the speedup of an algorithm increases roughly linearly with the number of processes for a fixed problem size, and the speedup curve for a strongly scalable parallel algorithm should therefore be close to a straight line. Strong scalability is difficult to achieve for a parallel algorithm, and, at best, an algorithm may be strongly scalable for a limited range of processes. Thus, if the problem size is kept constant, the efficiency will eventually begin to decrease when the number of processes exceeds the number of computational tasks to be distributed.

A more detailed analysis of the scalability of an algorithm that can reveal the rate at which the problem size must grow to maintain a constant efficiency requires an explicit functional form for $E(p, n)$. We will show an example of the derivation of an efficiency function in section 5.3.2. Additionally, it is important to clearly define the *problem size*. In computational complexity theory, the problem size is usually defined to be a measure for the size of the input required for an algorithm.[4] For example, the problem size for a matrix–matrix multiplication involving matrices of dimensions $m \times m$ would be m^2. This definition for the problem size is also consistent with the one usually employed in computational chemistry, where the problem size is defined as the size of the molecule being studied, expressed, for example, in terms of the number of atoms, electrons, or basis functions.

5.3 Performance Modeling

Having introduced various performance metrics, we are now ready to start developing performance models for parallel programs. A performance model for a parallel algorithm should be capable of predicting the execution time of the algorithm running on a given parallel computer as a function of the problem size and the number of processes employed. From the execution time, other performance measures, such as speedup and efficiency, can then be computed. We emphasize that very accurate prediction of the execution time is not a requirement for a performance model; indeed, for complex algorithms involving disparate computational tasks and communication operations, this would require undue effort. Rather, the performance model should simply provide an estimate for the execution time that is sufficiently accurate to capture the essential performance characteristics, such as scalability, communication bottlenecks, and degree of parallelism, that are required to assess the parallel performance of the algorithm.

5.3.1 Modeling the Execution Time

The execution time for a parallel algorithm is a function of the number of processes, p, and the problem size, n. Additionally, the execution time depends parametrically on several machine-specific parameters that characterize the communication network and the computation speed: the latency and the inverse of the bandwidth, α and β, respectively (both defined in section 5.1), and the floating point operation (flop) rate, γ, which is a measure for the time needed to carry out one floating point operation.

The values for these machine-specific parameters can be somewhat dependent on the application. For example, the flop rate can vary significantly depending on the type of operations performed. The accuracy of a performance model may be improved by using values for the machine-specific parameters that are obtained for the type of application in question, and the use of such empirical data can also simplify performance modeling. Thus, if specific, well-defined types of operations are to be performed in a parallel program (for instance, certain collective communication operations or specific computational tasks), simple test programs using these types of operations can be written to provide the appropriate values for the pertinent performance parameters. We will show examples of the determination of application specific values for α, β, and γ in section 5.3.2.

Let us develop an expression for the execution time for a parallel program, assuming that computation is not overlapped with communication and that no process is ever idle. Each process, then, will always be engaged in either computation or communication, and the execution time can be expressed as a sum of the computation and the communication times

$$t(p, n; \alpha, \beta, \gamma) = t_{\text{comp}}(p, n; \gamma) + t_{\text{comm}}(p, n; \alpha, \beta, \gamma) \tag{5.14}$$

where we use the notation $t(p, n; \alpha, \beta, \gamma)$ to indicate that t is a function of p and n and depends parametrically on α, β, and γ. To estimate the computational time, we need to know the total number of floating point operations required. We will assume that this flop count is independent of the number of processes, and we will represent it by the function $g(n)$. Assuming that the fraction f of the work is inherently sequential, and that the parallelizable fraction, $1 - f$, can be parallelized ideally, the computational time can be expressed as

$$t_{\text{comp}}(p, n; \gamma) = g(n)\left[f + \frac{(1-f)}{p}\right]\gamma. \tag{5.15}$$

The communication time required by an algorithm is a function of the number of messages sent and of the length of these messages. Using the idealized machine model (Eq. 5.1), the total communication time can be estimated as

$$t_{\text{comm}}(p, n; \alpha, \beta, \gamma) = m(p, n) \times \alpha + l(p, n) \times \beta + h(p, n) \times \gamma \tag{5.16}$$

where $m(p, n)$ is the number of messages sent by a process, $l(p, n)$ represents the combined length of all these messages, and $h(p, n)$ is the number of floating point operations required in the communication steps (some communication steps, such as the reduce operations discussed in section 3.2.3, require floating point operations to be performed). In general, both the number of messages and the amount of data to be communicated and processed depend on the number of processes as well as the problem size. If collective communication operations are employed, modeling of the communication time requires knowledge of the specific implementation used for the operation of interest. Performance models have been developed for most of the commonly used collective communication operations, and we have included examples of these in section 3.2. The communication overhead for a parallel program can sometimes be, at least partially, hidden by overlapping communication and computation. This, in effect, will reduce the values for α and β entering Eq. 5.16. To achieve such masking of the communication overhead, asynchronous message passing must be employed, and the communication scheme must be designed specifically to interleave communication and computation. We will illustrate algorithms that overlap communication and computation in sections 8.4 and 9.4.

The performance models we have discussed so far have assumed that equal amounts of work will be done on all processes and have not taken load imbalance into account. Load imbalance arises when different amounts of work are assigned to different processes, and for applications involving nonuniform computational tasks, load imbalance can be one of the major factors contributing to lowering the parallel efficiency. We have assumed previously (Eq. 5.14) that a process is always engaged in either computation or communication; when taking load imbalance into account, however, a process may also be idle, waiting for other processes to finish tasks, and the total execution time can be modeled as

$$t = t^i_{\text{comp}} + t^i_{\text{comm}} + t^i_{\text{idle}} \tag{5.17}$$

where the superscript i indicates timings for process P_i. The sum of t^i_{comp}, t^i_{comm}, and t^i_{idle} is the same for all processes, but the individual contributions vary across processes. Load imbalance is then measured as the difference between the total execution time and the average active time across processes. Defining active time as $t^i_{\text{comp}} + t^i_{\text{comm}}$, the load imbalance can be expressed as

$$\text{Load imbalance} = t - \frac{\sum_{i=1}^{p}(t^i_{\text{comp}} + t^i_{\text{comm}})}{p} \tag{5.18}$$

where t is the total execution time (Eq. 5.17), and p represents the number of processes.

The presence of nonuniform computational tasks whose sizes are not known in advance is often the cause of load imbalance, and quantitative modeling of load imbalance can therefore be difficult to do. However, simulations that involve distribution of nonuniform tasks can sometimes provide empirical data that can be used to model load imbalance. In chapter 7 we will illustrate the use of empirical data to model load imbalance in the computation of the two-electron integrals, which is a required step in many quantum chemical methods. Parallel programs involving uniform computational tasks will experience load imbalance whenever the number of tasks is not a multiple of the number of processes. This kind of load imbalance is easier to include in a performance model because the amount of work assigned to the process

```
/* Loop over i values to be processed locally */
For i = this_proc×n/p, i < (this_proc+1)×n/p, i = i + 1
    c_i = A_i b        (form the dot product of i'th row of A with b)
End for
All-to-all broadcast of c        (put entire c vector on all processes)
```

FIGURE 5.4
Outline of a simple parallel algorithm to perform the matrix–vector multiplication $\mathbf{Ab} = \mathbf{c}$. $\mathbf{A}$ is an $n \times n$ matrix distributed by rows, $\mathbf{b}$ and $\mathbf{c}$ are replicated vectors, p is the number of processes, and this_proc is the process ID. A static work distribution scheme is used, and a single global communication operation is required. Each process computes the elements c_i of the $\mathbf{c}$ vector corresponding to the locally stored rows of $\mathbf{A}$, and an all-to-all broadcast operation puts a copy of the entire $\mathbf{c}$ vector on all processes.

with the heaviest load can be predicted, and the corresponding execution time can be used in the performance model.

5.3.2 Performance Model Example: Matrix-Vector Multiplication

To illustrate a performance model for a simple parallel program, consider the parallel matrix–vector multiplication program outlined in Figure 5.4. The program performs the multiplication of the matrix $\mathbf{A}$ with the vector $\mathbf{b}$, producing the product vector $\mathbf{c}$; $\mathbf{A}$ is distributed by rows across processes, and $\mathbf{b}$ and $\mathbf{c}$ are replicated. The multiplication is parallelized by letting each process compute the product of the locally held rows of $\mathbf{A}$ with $\mathbf{b}$. Every process computes its part of the product as a series of vector–vector dot products, processing one row of $\mathbf{A}$ at a time, each dot product yielding the corresponding element of the vector $\mathbf{c}$. When all processes have completed the multiplication step, a global communication operation (an all-to-all broadcast) is performed to put a copy of the entire $\mathbf{c}$ vector on all processes.

Let us develop a performance model for the parallel matrix–vector multiplication. We first note that if the dimensions of $\mathbf{A}$ are $n \times n$, the maximum number of processes that can be utilized in this parallel algorithm equals n. The total number of floating point operations required is n^2 (where we have counted a combined multiply and add as a single operation), and provided that the work is distributed evenly, which is a good approximation if $n \gg p$, the computation time per process is

$$t_{\text{comp}} = \frac{n^2}{p}\gamma \tag{5.19}$$

where γ is the floating point operation rate, and p is the number of processes. The only communication step required is the all-to-all broadcast of $\mathbf{c}$. To perform this operation we will use an algorithm based on a binomial tree all-reduce operation for which the communication time can be modeled as (Eq. 3.7)

$$t_{\text{comm}} = \log_2 p[2\alpha + n(2\beta + \gamma)]. \tag{5.20}$$

We will use this algorithm instead of the all-to-all broadcast algorithm provided in the employed implementation of MPI because the latter algorithm displayed very irregular performance.‡ A performance model for a matrix–vector multiplication that uses the all-to-all broadcast is discussed in section 6.4.1. The total execution time, the speedup, and the efficiency can then be expresssed as the following functions of p and n

$$t_{\text{total}}(p, n) = \frac{n^2}{p}\gamma + \log_2 p[2\alpha + n(2\beta + \gamma)] \tag{5.21}$$

$$S(p, n) = \frac{t_{\text{total}}(n, 1)}{t_{\text{total}}(n, p)} = \frac{p}{1 + p\log_2 p[2\alpha + n(2\beta + \gamma)]/(n^2\gamma)} \tag{5.22}$$

$$E(p, n) = \frac{1}{1 + p\log_2 p[2\alpha + n(2\beta + \gamma)]/(n^2\gamma)}. \tag{5.23}$$

Several performance features of the algorithm can be gleaned from the above equations without determining the values of α, β, and γ. Thus, it is apparent that the algorithm is not strongly scalable, because the efficiency, $E(p, n)$, is a monotonically decreasing function of p for any fixed problem size. The algorithm is weakly scalable, however, because the efficiency may be kept nearly constant as the number of processes increases provided that the problem size increases as well. To determine how rapidly the problem size must grow, consider the term $p\log_2 p[2\alpha + n(2\beta + \gamma)]/(n^2\gamma)$ from the denominator of Eq. 5.23. The value of this expression must not increase as p increases, and n must therefore grow at approximately the same rate as $p\log_2 p$. If we define the problem size as n^2, in keeping with the definition introduced in section 5.2.2, the problem size must then increase as $(p\log_2 p)^2$ to maintain a nearly constant efficiency as the number of processes increases.

From Eq. 5.22 we can also compute the the maximum attainable speedup for the algorithm. If we differentiate the expression for the speedup with respect to p, keeping n fixed and using the definition $A = [2\alpha + n(2\beta + \gamma)]/(n^2\gamma \ln 2)$, we get

$$\frac{\partial S(p, n)}{\partial p} = \frac{1 - pA}{(1 + Ap\ln p)^2}. \tag{5.24}$$

Therefore, the maximum speedup will be obtained when the number of processes is $p_{\text{max}} = 1/A = n^2\gamma \ln 2/[2\alpha + n(2\beta + \gamma)]$. For process counts $p < p_{\text{max}}$, $S(p, n)$ is an increasing function of p, and the maximum speedup that can be obtained with the algorithm is $S_{\text{max}} = S(p_{\text{max}}, n)$; p_{max} is also an upper limit for the number of processes that should be used with this algorithm because increasing p beyond p_{max} will lower the speedup.

‡ A given implementation of MPI often will not provide optimal performance for all MPI operations; to ascertain that the MPI implementation delivers reasonable and predictable performance, the user should therefore compare the actual performance with that predicted by the appropriate performance models.

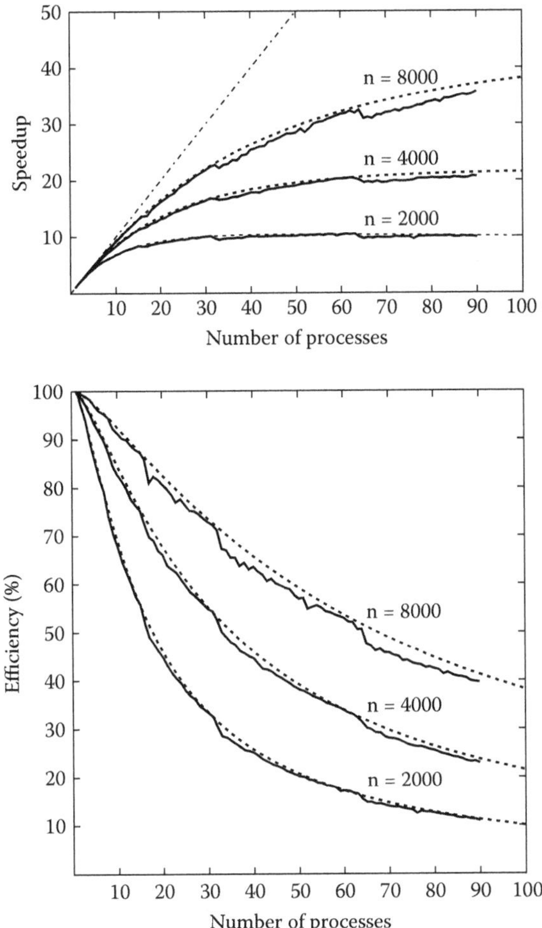

FIGURE 5.5
Predicted and measured speedups and efficiencies for the simple parallel matrix–vector multiplication outlined in Figure 5.4. Dashed curves represent predictions by the performance model, solid curves show measured values, and the dot-dashed line is the ideal speedup curve. The matrix dimensions are $n \times n$.

To determine values for the machine parameters α, β, and γ, a series of test runs were performed using a Linux cluster.[5] The value for γ was estimated to be 4.3 ns by timing single-process matrix–vector multiplications for various matrix sizes. To model the communication time, the values of α and the sum $2\beta + \gamma$ are required; these values were found to be $\alpha = 43$ μs and $2\beta + \gamma = 72$ ns/word (using 8 byte words) by timing the all-reduce operation as a function of the number of processes for a number of problem sizes and fitting the data to a function of the form of Eq. 5.20. Using these values for the machine parameters, the performance model was used to predict the speedups and efficiencies that can be obtained for the parallel matrix–vector multiplication, and the results are shown in Figure 5.5 along with the speedups and efficiencies measured by running the program. From these plots it is evident that the performance model works very well for the range of problem sizes and process counts investigated. The expression used to model the communication time for the all-reduce operation is strictly correct only when the number of processes is a power of two, and, otherwise, underestimates the communication time. Consequently, the speedups predicted by the model are slightly higher than the observed speedups for process counts that are not a power of two, and this effect is more pronounced for large problem sizes. For instance, for $n = 8000$, the observed speedups drop below the predicted values just after $p = 64$, and as p increases toward the next power of two, the observed speedups approach the predicted values again. The parallel matrix–vector multiplication is an application involving only uniform computational tasks and a single type of collective communication, and it is therefore relatively easy to provide an accurate performance model. For this application, there are n computational tasks, and load imbalance will be essentially negligible when $n \gg p$. In chapter 7 we will illustrate how to model load imbalance for an application with unevenly sized computational tasks.

The speedup curves in Figure 5.5 also clearly demonstrate that the parallel efficiency is a rapidly decreasing function of p. The decrease in the efficiency is caused by the large communication overhead involved in the all-reduce operation, which becomes a bottleneck that severely limits the speedup. Using the expression derived for the process count, $p_{\max}$, that yields the maximum speedup, we find the maximum attainable speedups with the algorithm for $n = 2000$, 4000, and 8000 to be roughly 10, 22, and 43 respectively, obtained at the corresponding $p_{\max}$ values of 51, 126, and 284. The weak scalability of the algorithm is also demonstrated by the plot, which displays higher speedups for larger problem sizes for a given process count. The performance model predicts that increasing the dimension n at roughly the same rate as $p \log_2 p$ should yield a constant efficiency as p increases, and this is consistent with the measured efficiencies. Finally, we note that performance models for parallel matrix–vector multiplication using a row-distributed matrix as well as a block-distributed matrix will be discussed in section 6.4, demonstrating the higher scalability of the block-distributed approach.

5.4 Presenting and Evaluating Performance Data: A Few Caveats

In the previous sections of this chapter we discussed how to do performance modeling for parallel programs, and we will here briefly consider a few important points to keep in mind when presenting performance data for a parallel algorithm or evaluating performance data reported in the literature.

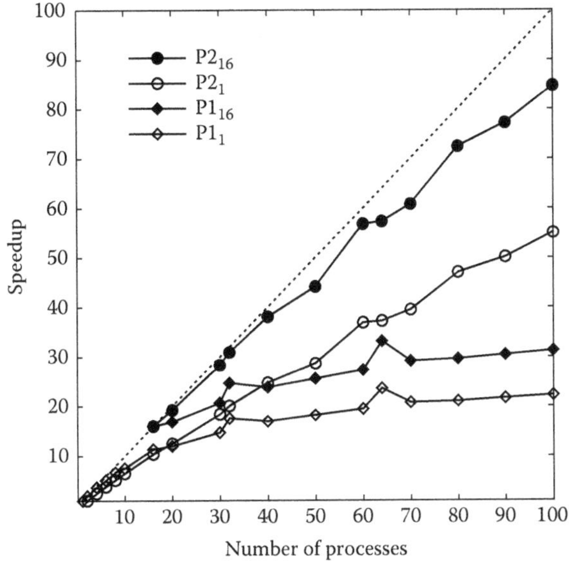

FIGURE 5.6
Speedups for MP2 algorithms P1 and P2 (dynamic version) measured relative to timings for one process ($P1_1$ and $P2_1$) and sixteen processes ($P1_{16}$ and $P2_{16}$). Computations were performed on a Linux cluster[12] for the uracil dimer molecule using the cc-pVDZ basis set (cf. Figure 9.8). Inflated speedup curves are obtained by measuring speedups relative to a number of processes greater than one.

A number of ways to report misleading parallel performance data have been discussed elsewhere,[6,7] including how to boost performance data by comparing with code that is nonoptimal in a number of ways. Performance data are most often presented in the form of speedup curves, and it is therefore important to ascertain that the presented speedups are, in fact, representative of the typical parallel performance of the algorithm. Below we will discuss a couple of commonly encountered practices for presenting speedups that can lead to misrepresentation of performance data.

In Figure 5.6, speedup curves are shown for two algorithms computing energies with second order Møller–Plesset perturbation theory. The algorithms, designated P1 and P2, are explained in detail in chapter 9. For each algorithm, two speedup curves were produced from one set of timings: one curve measures speedups relative to the computational time on a single process, and the other curve measures speedups relative to timings obtained on sixteen processes, setting the speedup $S(p = 16)$ equal to 16. Reporting speedups in this way, namely, measuring speedups relative to a number of processes $p_{\min}$ and assuming that $S(p_{\min}) = p_{\min}$, is a fairly common practice, in part justified if $p_{\min}$ is the smallest number of processes that will allow the calculation to run. However, if the assumption $S(p_{\min}) = p_{\min}$ is unwarranted, this practice can produce inflated speedup curves. For the speedup curves

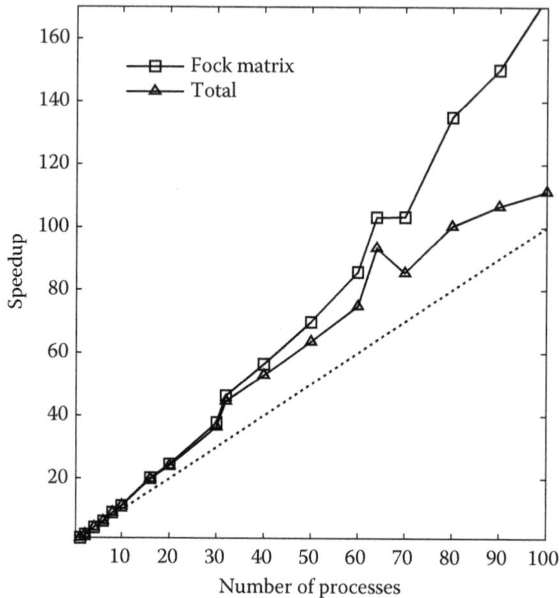

FIGURE 5.7
Superlinear speedups for Fock matrix formation in the iterative part of the Hartree–Fock procedure as a consequence of storing a larger fraction of the integrals on each process as the number of processes increases. Speedups for the entire Hartree–Fock procedure are shown as well. Computations were performed on a Linux cluster[12] for the uracil dimer using the aug-cc-pVTZ basis set (cf. Figure 8.3). A static task distribution of atom quartets was employed (see section 8.3 for details of the algorithm).

shown in Figure 5.6 there is a nonnegligible performance degradation when going from one to sixteen processes; the speedup curves measured relative to sixteen processes, therefore, are somewhat misrepresentative, exaggerating the parallel performance of the algorithms. We note that on parallel computers consisting of multiprocessor nodes, it may be reasonable to measure the speedup as a function of the number of nodes, rather than processors, because a node, not a single processor, is the repeating unit.

We have seen in section 5.2 that the ideal speedup for a process count of p is $S(p) = p$, but that superlinear speedups are possible for some algorithms. For instance, integral-direct algorithms in quantum chemistry can take advantage of the increase in aggregate memory as the number of processes increases to achieve superlinear speedups. In Figures 5.7 and 5.8 we show examples of superlinear speedup curves for two integral-direct quantum chemistry algorithms. Figure 5.7 displays speedups obtained with a direct Hartree–Fock program using replicated data, illustrating speedups both for the entire Hartree–Fock procedure and for the computationally dominant step in the procedure, the formation of the Fock matrix. The Hartree–Fock procedure is iterative, and the Fock matrix must be formed in each iteration; formation of

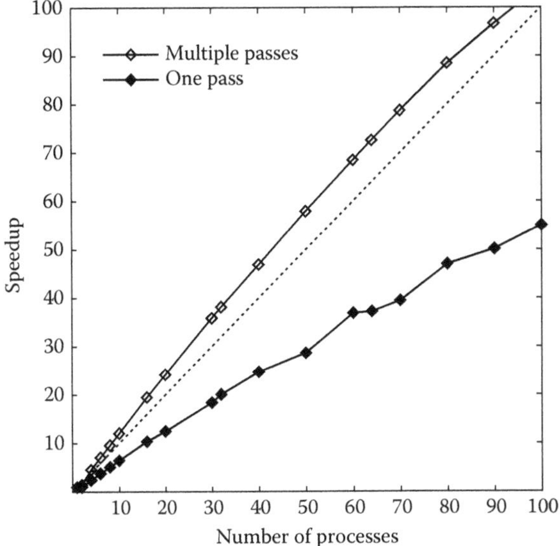

FIGURE 5.8
Superlinear speedups for an MP2 calculation requiring multiple integral passes, using the dynamic version of the P2 algorithm. The corresponding speedups when doing only one integral pass are shown as well, and the dashed line represents linear speedups. Data were obtained on a Linux cluster[12] for the uracil dimer molecule using the cc-pVDZ basis set (cf. Figure 9.8).

the Fock matrix requires the two-electron integrals, which therefore must be computed in every iteration. A certain number of the integrals, however, do not need to be recomputed because they can be stored in memory, and as the aggregate memory increases with the number of processes, a smaller fraction of the integrals must be recomputed. This results in superlinear speedups, producing a concave upward speedup curve (see section 8.3). The superlinear speedups in the Fock matrix formation step also cause the entire Hartree–Fock procedure to display superlinear speedups, although the performance degradation observed in various steps of the procedure reduces the slope of the speedup curve for the total procedure.

The speedup curves in Figure 5.8 were obtained for the computation of energies with second-order Møller–Plesset perturbation theory using the integral-direct algorithm P2 explained in detail in section 9.4. The algorithm can achieve superlinear speedups by utilizing the increased aggregate memory to reduce the number of integral passes, and, hence, the total computational work, as the number of processes increases. If enough memory is available that only one integral pass is required when $p = 1$, however, the algorithm will not display superlinear speedups. The speedups in Figure 5.8 were measured for two series of runs that were identical except for the amount of memory allocated per process: in the first case enough memory was allocated per process that only one integral pass was required; in the second case, less memory was allocated per process so that the required number of integral passes was 4, 2, and 1 for $p = 1$, $p = 2$, and $p \geq 4$, respectively. While the superlinear speedup curve does represent actual speedups, it masks the fact that there is a significant performance degradation as the number of processes increases. The other speedup curve, however, representing speedups obtained when the total amount of computation is constant for all process counts, gives a clear picture of how well the algorithm has been parallelized. Although superlinear speedups may be legitimate, they should not be presented without investigating how they arise. For the purpose of performance analysis, the steps that display superlinear speedups should be identified and, if possible, it may be informative to show the speedups for these and other steps separately.

5.5 Further Reading

We have used expressions involving the latency, α, and inverse bandwidth, β, to model the communication time. An alternative model, the Hockney model,[8,9] is sometimes used for the communication time in a parallel algorithm. The Hockney model expresses the time required to send a message between two processes in terms of the parameters r_∞ and $n_{\frac{1}{2}}$, which represent the asymptotic bandwidth and the message length for which half of the asymptotic bandwidth is attained, respectively. Metrics other than the speedup and efficiency are used in parallel computing. One such metric is the Karp–Flatt metric,[10] also referred to as the experimentally determined serial fraction. This metric is intended to be used in addition to the speedup and efficiency, and it is easily computed. The Karp–Flatt metric can provide information on parallel performance characteristics that cannot be obtained from the speedup and efficiency, for instance, whether degrading parallel performance is caused by incomplete parallelization or by other factors such as load imbalance and communication overhead.[11]

References

1. Benchmark programs for determining unidirectional and bidirectional bandwidths are available on the World Wide Web at http://mvapich.cse.ohio-state.edu. The Intel MPI Benchmarks (formerly known as Pallas MPI Benchmarks) can be found on the Intel web site http://www.intel.com.
2. Amdahl, G. M. Validity of the single-processor approach to achieving large scale computing capabilities. *AFIPS Conf. Proc.* 30:483–485, 1967.
3. Gustafson, J. L. Reevaluating Amdahl's law. *Commun. ACM* 31:532–533, 1988.
4. Garey, M. R., and D. S. Johnson. *Computers and Intractability: A Guide to the Theory of NP-Completeness*, chapter 1. New York: W. H. Freeman and Company, 1979.

5. A Linux® cluster consisting of nodes with two single-core 3.06 GHz Intel® Xeon® processors (each with 512 KiB of L2 cache) connected via a 4x Single Data Rate InfiniBand network with a full fat tree topology.
6. Bailey, D. H. Twelve ways to fool the masses when giving performance results on parallel computers. *Supercomputing Review*, pp. 54–55, June 11, 1991.
7. Bailey, D. H. Misleading performance reporting in the supercomputing field. *Sci. Prog.* 1:141–151, 1992.
8. Hockney, R. W., and C. R. Jesshope. *Parallel Computers 2: Architecture, Programming and Algorithms*. Bristol: IOP Publishing/Adam Hilger, 1988.
9. Hockney, R. W. The communication challenge for MPP: Intel Paragon and Meiko CS-2. *Parallel Comput.* 20:389–398, 1994.
10. Karp, A. H., and H. P. Flatt. Measuring parallel processor performance. *Commun. ACM* 33:539–543, 1990.
11. Quinn, M. J. *Parallel Programming in C with MPI and OpenMP*. New York: McGraw-Hill, 2003.
12. A Linux® cluster consisting of nodes with two single-core 3.6 GHz Intel® Xeon® processors (each with 2 MiB of L2 cache) connected with a 4x Single Data Rate InfiniBand™ network using Mellanox Technologies MT25208 InfiniHost™ III Ex host adaptors and a Topspin 540 switch. The InfiniBand host adaptors were resident in a PCI Express 4x slot, reducing actual performance somewhat compared to using them in an 8x slot. MPICH2 1.0.5p4 was used with TCP/IP over InfiniBand IPoIB because this configuration provided the thread-safety required for some applications.

6

Parallel Program Design

Before undertaking the task of writing a parallel program, which may require a significant amount of effort, it is important to consider what one wishes to gain from the parallel implementation. Is a shorter time to solution the only requirement? Or should the parallel program also be able to tackle larger problem sizes, use a large number of processes, and run with high parallel efficiency to avoid wasting computing resources? In general, parallel execution of a given computational problem can be achieved in multiple ways, and the optimal solution may depend on factors such as the range of problem sizes that will be targeted, the number of processes that will be used, the amount of memory available, and the performance characteristics of the communication network. It is therefore important to take these factors into consideration in the design phase so that the parallel algorithm can be tailored for high performance under the circumstances likely to be encountered in practice. The design process may involve prioritizing performance characteristics, such as whether to minimize the operation count, the memory requirement, or the communication overhead, and then making tradeoffs based on the priorities. For instance, it may be possible to reduce the memory requirement or the communication overhead by allowing some redundant computation.

The first step in the development of a parallel algorithm usually involves determining how the computational work can be partitioned into smaller tasks that can be processed concurrently and how to distribute the data involved in the computation. One may distinguish between parallelization schemes driven by the distribution of data, so-called *data decomposition* (also known as *domain decomposition*), and strategies centered on a partitioning of computational work, known as *functional decomposition*. In data decomposition schemes, one first works out a partitioning of the data associated with a problem, and this data distribution, in turn, determines the distribution of the associated computational tasks. A functional decomposition, on the other hand, first determines how to distribute the computational work, and a data distribution appropriate for this work partitioning is subsequently worked out. These complementary parallelization strategies have been discussed in the literature,[1–3] and we will not here focus on which distribution drives the parallelization; in practice, parallel algorithm design in quantum chemistry usually involves simultaneous consideration of both task and data distributions.

In the following, we will address the major issues involved in the design of a parallel algorithm, including the partitioning of the problem into smaller tasks that can be performed in parallel, the distribution of data, and the development of a communication scheme to handle the required exchange of data between processes. We will also examine techniques for improving the parallel performance, including load balancing strategies and ways to reduce the communication overhead.

6.1 Distribution of Work

A key point in the design of a parallel algorithm is the partitioning of the computational work into a set of smaller tasks that can be performed concurrently. Let us first introduce a couple of terms pertaining to the distribution of work that will be useful in discussing different parallelization schemes. The *degree of parallelism* (also called the *degree of concurrency*) of a parallel algorithm is defined as the number of tasks that can be executed simultaneously. Often, different parts of a parallel algorithm will have different degrees of parallelism, and it is therefore reasonable to use the term degree of parallelism as a measure for the maximum number of processes that can be used efficiently in the execution of a given algorithm. The *granularity* of a parallel algorithm refers to the size of the individual computational tasks that are distributed among processes (or threads in a hybrid approach), where the *task size* is a measure for the time required to complete the task. The concept of granularity is illustrated in Figure 6.1: if the tasks are small, the algorithm is said to be *fine-grained*; a *coarse-grained* algorithm, on the other hand, employs relatively large tasks. Generally, a fine-grained algorithm involves more overhead in terms of communication and synchronization of processes and can suffer from low efficiency because the work is divided into small chunks. However,

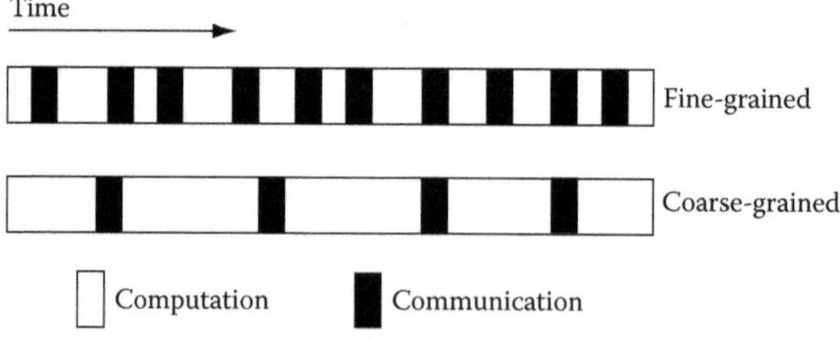

FIGURE 6.1
Interspersion of computation and communication time on a process in a fine-grained and a coarse-grained parallel algorithm.

fine-grained parallelism may provide a better opportunity for evenly balancing the load because there are more tasks to be distributed. Coarse-grained parallelism allows individual processes to work independently for longer stretches of time, and it tends to involve less communication overhead and to be easier to implement; the disadvantage of a coarse-grained algorithm is a smaller degree of parallelism, which increases the likelihood of load imbalance and limits the number of processes that can be used efficiently. As a general strategy for determining the appropriate level of granularity for an algorithm, task sizes should be chosen so as to allow an even distribution of work (within the desired range of process counts), which is a prerequisite for achieving high parallel efficiency.

Depending on the desired degree of parallelism for an algorithm, it may be advantageous to first explore relatively fine-grained work partitioning schemes because the availability of more tasks allows more flexibility in the parallel design. If necessary, individual tasks, subsequently, can be grouped together into larger chunks to form a more coarse-grained algorithm. When deciding on a work distribution scheme, it is necessary to consider both the number and sizes as well as the uniformity of the tasks and also possible interdependencies between tasks that may impose restrictions on the order in which they are carried out. In the following we will discuss two schemes for distributing the work, namely *static* and *dynamic task distributions*. A static scheme uses a predetermined distribution of the work that cannot be altered during the computation, whereas a dynamic load balancing approach determines the work distribution during the execution of the program, assigning work to processes as they become idle until there are no tasks left.

6.1.1 Static Task Distribution

The most straightforward way to partition the work is to use a static task distribution, assigning tasks to processes according to a predetermined scheme. With a static task distribution, work cannot be redistributed during the execution of the program if load imbalance arises, and static distributions therefore are well-suited mostly for computational problems involving uniform tasks requiring the same execution time. Static distribution schemes offer a number of advantages relative to dynamic schemes. Firstly, the static schemes tend to be easier to implement, and they usually do not require any communication to distribute the work. Moreover, for a static task distribution it can be determined in advance which tasks will be assigned to a given process, and this may provide the opportunity to store data locally on the processes where it will be needed.

Example 6.1 Exploiting Data Locality in a Parallel Direct Hartree–Fock Program

If we can determine in advance which tasks will be assigned to a given process, we can also predetermine what data that process will need to access. Hence, in some cases it may be possible to use a data distribution

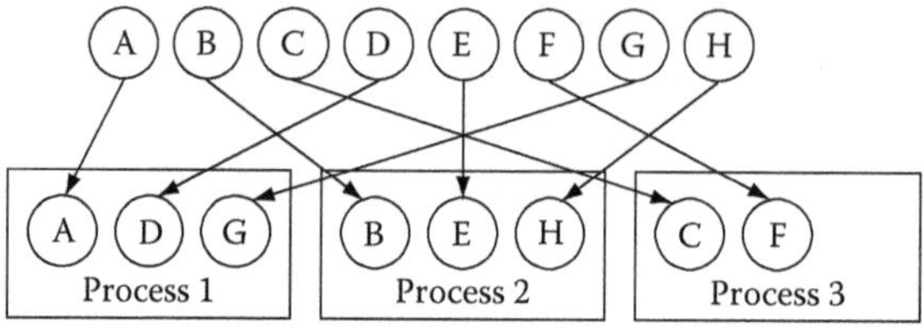

FIGURE 6.2
Distribution of tasks (labeled A–H) across three processes using a round-robin distribution.

scheme that simply places the data on the processes where it will subsequently be needed. This approach can reduce the communication requirement and may also reduce the computational cost. Consider, for instance, the computation of the Fock matrix, a computationally dominant step in the Hartree–Fock procedure (see chapter 8). The Hartree–Fock procedure is iterative, and the Fock matrix must be computed in each iteration. Computation of the Fock matrix involves a contribution from the two-electron integrals (Eq. 8.5) of the form

$$F_{\mu\nu} = F_{\mu\nu} + D_{\rho\lambda}(\mu\nu|\rho\lambda) \tag{6.1}$$

where $D_{\rho\lambda}$ represents an element of the density matrix, $(\mu\nu|\rho\lambda)$ is a two-electron integral, and μ, ν, ρ, and λ denote atomic orbitals. If a direct approach is used in which the two-electron integrals are computed when they are needed, the integrals must be computed in each iteration. Work is distributed by letting each process compute a subset of the $(\mu\nu|\rho\lambda)$ integrals, and a given process thus computes a subset of the integrals in every iteration. For a static work distribution scheme, however, the same integrals will be required by a given process in each iteration, and this process can therefore store a subset of these integrals (depending on the available memory) and reuse them, saving computational time. If a dynamic work distribution scheme were employed, a process most likely would need a different set of integrals in every iteration, and reusing locally stored integrals would not be possible.

6.1.1.1 *Round-Robin and Recursive Task Distributions*

Commonly used static task distribution schemes include round-robin and recursive distributions. A round-robin work distribution scheme is illustrated in Figure 6.2. In this type of scheme, work is distributed by looping through a task list, assigning one task to each process in turn and repeating until all tasks have been assigned. A simple round-robin allocation of tasks can be implemented as outlined in Figure 6.3. If the tasks are not uniform, load balance can usually be improved by first randomizing the tasks in the task list; round-robin scheduling using such randomization of tasks can balance the work reasonably well even for quite nonuniform task sizes provided that

```
For i = this_proc, i < n_task, i = i + p
    Process task i
End for
```

FIGURE 6.3
Outline of a simple round-robin static task distribution. The number of tasks is n_{task}, p is the number of processes, and this_proc is the process ID.

the number of tasks is much larger than the number of processes. In Part II of this book, a round-robin distribution is used for most of the algorithms that employ a static task distribution.

A recursive work distribution scheme involves recursive subdivisions of the computational problem to create smaller tasks that can be solved concurrently by individual processes. Recursive work distribution schemes are typically used for computational problems that lend themselves to a divide-and-conquer parallelization approach. For instance, finding the maximum value in an unsorted array and sorting the elements in an array are problems that can be solved using recursive algorithms.

Example 6.2 Parallel Quicksort

The widely used quicksort algorithm is a recursive algorithm for sorting a sequence of numbers.[4] It uses a divide-and-conquer strategy to divide the sequence of numbers into successively smaller subsequences until the subsequences are of length one, or, at least, sufficiently short that their sorting is a trivial problem. An outline of the quicksort algorithm is shown in Figure 6.4. Initially, an element from the sequence is selected as a pivot, and the sequence is then divided into two smaller sequences, one containing the elements smaller than or equal to the pivot, the other containing the elements greater than the pivot. These two subsequences are then sorted by recursive application of quicksort. In the parallel version of the quicksort procedure, the recursive calls to quicksort (`Quicksort(Left,nleft)` and `Quicksort(Right,nright)`) are performed in parallel. This can be done by treating these calls as new tasks to be started up in separate processes. The recursive subdivision of tasks and their assignment to new processes are illustrated in Figure 6.5. To achieve an efficient parallel quicksort, it is necessary also to parallelize the creation of the left and right subsequences, for instance, by letting each process be responsible for a subset of the original sequence and assigning the elements in this subset to the left and right subsequences. Although the recursive parallel quicksort algorithm may be conceptually simple, the implementation of an efficient parallel quicksort algorithm is somewhat involved and is beyond the scope of this text. A more thorough discussion of parallel implementations of quicksort can be found elsewhere.[2] Note that the parallel quicksort algorithm illustrated here is an example of a static work distribution scheme that, unlike most static schemes, requires communication to distribute the tasks.

```
Quicksort(A,n) {
    If n eq 1: return A[0]

    int q = int(n*rand())
    int pivot = A[q]

    double Left[n], Right[n]
    int nleft = 0, nright = 0

    For i = 0, i < n, i = i + 1
        If A[i] ≤ pivot
            Left[nleft] = A[i]
            nleft = nleft + 1
        Endif
        If A[i] > pivot
            Right[nright] = A[i]
            nright = nright + 1
        Endif
    End for

    Return concatenate(Quicksort(Left,nleft),Quicksort(Right,nright)) }
```

FIGURE 6.4
Outline of a sequential quicksort algorithm for sorting the n elements in the array **A**. The pivot is here chosen randomly among the elements of **A**. The Left array holds the elements smaller than or equal to the pivot, and the Right array contains elements greater than the pivot. The quicksort function is called recursively via the calls to `Quicksort(Left,nleft)` and `Quicksort(Right,nright)`, and the resulting arrays are concatenated before returning.

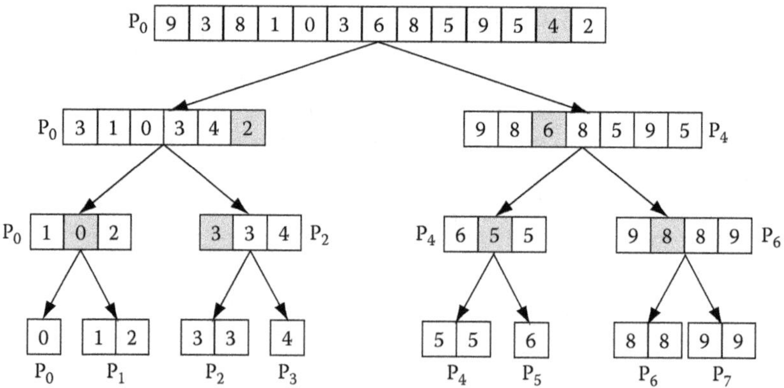

FIGURE 6.5
Recursive subdivision of tasks in parallel quicksort using eight processes, P_0–P_7. The array to be sorted is subdivided by recursive calls to quicksort, indicated by arrows representing assignment of work to a new process. The pivot is shown in the gray box.

6.1.2 Dynamic Task Distribution

Dynamic task distribution is a load balancing strategy particularly useful for computational problems involving nonuniform tasks. In this type of parallelization, the task distribution is determined during the execution of the program by assigning work to processes as soon as they become idle. A dynamic task distribution, therefore, responds to arising load imbalance during the execution of a program by allocating work only to the processes that are ready for new tasks and not assigning further work to those processes that are already engaged in computation. Dynamic task distribution tends to be more difficult to implement than static distribution schemes because it may involve nontrivial communication patterns, and, as noted in section 6.1.1, it may not be possible to exploit data locality to the same extent in a dynamic task distribution model because the dynamic assignment of tasks to processes may necessitate redistribution of data. Dynamic task distribution, however, usually offers improved load balance relative to static schemes and allows efficient use of more processes. Dynamic load balancing schemes can be implemented by means of a manager–worker approach with one process in charge of the distribution of work to others, or by using a decentralized work scheduling mechanism in which all processes exchange tasks with each other as necessary to balance the work. We will discuss these two approaches next.

6.1.2.1 Manager–Worker Model

In a *manager–worker model* (also known as a *master–slave model*) one process, the manager, distributes work to the other processes, the workers. A dynamic manager–worker task distribution scheme is outlined in Figure 6.6. In this scheme, the manager keeps a list of the computational tasks, and upon request the manager will send the next available task to a worker process. Each worker will process one task at a time and, when finished, request a new task from the manager. This continues until there are no tasks left, at which point a task request from a worker will be answered by a message from the manager informing the worker that all tasks have been completed. The parallel algorithms using dynamic load balancing in Part II of this book all employ a manager–worker model.

Dynamic manager–worker distribution schemes are most easily implemented by means of blocking send and receive operations, but the use of non-blocking operations will allow prefetching of task requests (on the manager) and tasks (on the workers). Such prefetching may reduce the amount of time workers spend waiting for task requests to be returned from the manager, and this can lead to improved parallel efficiency when the waiting time is nonnegligible, for instance, if there are numerous very small computational tasks. In the manager–worker model illustrated in Figure 6.6, the manager is dedicated to distributing tasks to workers and does not itself do any computation. Consequently, if there are p processes, only $p - 1$ processes are available to process the computational tasks, and the maximum parallel efficiency that can be obtained with this scheme is therefore $[(p-1)/p] \times 100\%$.

```
If (this_proc eq manager)
    Create task list
    int requests_remaining = n_task + p − 1
    int task_index = 0
    While (requests_remaining)
        Receive work request from a worker
        If (task_index < n_task)
            Send next task from list to worker
            task_index = task_index + 1
        Else
            Send "finished" message to worker
        Endif
        requests_remaining = requests_remaining − 1
    End while
Else
    int finished = 0
    Send request for new task to manager
    Receive task or "finished" from manager
    While (!finished)
        Process current task
        Send request for task to manager
        Receive new task or "finished" from manager
    End while
Endif
```

FIGURE 6.6
Outline of a dynamic manager–worker task distribution scheme. The number of tasks and processes are designated n_{task} and p, respectively. One process is assigned to be the manager, and the remaining $p - 1$ processes are the workers.

This limitation can be avoided by letting the manager participate in computation so that all p processes will process computational tasks, but the manager will then be less likely to be available to respond instantly to requests from the workers. If a manager–worker scheme involves a very large number of tasks or processes, the processing of requests for tasks on the manager may become a bottleneck. Thus, if the time required to finish a task on a worker is t_{task}, and the time required to process a request for a task is t_{req}, then t_{task}/t_{req} requests can be processed on the manager in the time it takes a worker to finish a task, and t_{task}/t_{req} therefore is the maximum number of worker processes that the manager can support. To avoid a bottleneck on the manager, the number of tasks can be reduced by grouping several tasks together, effectively increasing the granularity, although this approach can also make load balancing more difficult. Alternatively, a decentralized distribution of tasks can be employed as explained below.

6.1.2.2 Decentralized Task Distribution

A dynamic distribution of tasks can be carried out using a decentralized model that does not involve a manager but requires all processes to be engaged in computation and to exchange tasks with each other as processes become idle. This scheme can be implemented as follows. Initially, all tasks are distributed among processes using a simple predetermined scheme, and each process starts working on the tasks that were assigned to it. Once a process runs out of tasks, it will request tasks from other processes. For instance, when process P_i becomes idle, it will request tasks from P_j, which will then send a task back to P_i. It is common to use a work splitting strategy in which the process that has become idle receives half of the tasks remaining on another process. The exchange of work between processes continues until all tasks have been processed. If P_i requests work from P_j, and P_j does not have any tasks remaining, P_i will proceed to request a task from the next process and so on until a process responds with a task or all processes have been polled. Using this type of task distribution, it is not trivial to determine when all processes have finished, and termination detection algorithms may have to be used for this purpose (see section 6.6).

6.2 Distribution of Data

For computational problems involving large data structures, a well-designed data distribution scheme is a prerequisite for an efficient parallel implementation. In general, data must be distributed to take advantage of the increasing aggregate memory as the number of processes increases so that memory bottlenecks can be avoided when tackling larger problem sizes. Complex computational problems typically involve many data structures of different sizes, including input and output data and intermediates arising in the computation. Often, it may be possible to replicate small data structures without performance penalty, and this can simplify the parallel program design and reduce the need for interprocess communication.

Example 6.3 Distribution of Data in Quantum Chemistry Applications

Some data structures commonly encountered in quantum chemistry applications are listed in Table 6.1. For a problem size of n (where n is the number of basis functions or the number of atoms), there could be data structures whose storage requirements grow as $O(n)$, $O(n^2)$, $O(n^3)$, and $O(n^4)$. If arrays of $O(n^4)$ are stored, they must be distributed or serious

TABLE 6.1

Some data structures commonly encountered in quantum chemistry methods. The problem size (that is, the number of atoms or the number of basis functions) is denoted n

Size	Data Structures
$O(n)$	basis set, molecular coordinates, molecular orbital energies
$O(n^2)$	Fock matrix, density matrix, single-substitution amplitudes
$O(n^3)$	subsets of triple-substitution amplitudes
$O(n^4)$	two-electron integrals, double-substitution amplitudes

storage bottlenecks will result. Data structures of $O(n)$ are generally not distributed because their replication greatly simplifies programming and does not create performance problems. Whether to distribute structures of size $O(n^2)$ may depend on the method in question. If the $O(n^2)$ structures are the largest data structures to be stored, for instance, in direct Hartree–Fock and density functional theory, they are sometimes distributed, especially if large molecules are to be investigated. Otherwise, the $O(n^2)$ arrays are often replicated, in particular in correlated electronic structure methods where much larger data structures must also be handled. Intermediates requiring $O(n^3)$ storage arise, for example, in the computation of the perturbative triples contribution to the coupled-cluster energy, and these intermediates are not distributed because the bottleneck for the calculation usually is the $O(n^7)$ time requirement.

When designing a data distribution scheme, careful attention must be paid to creating a data layout that prevents storage bottlenecks and avoids excessive data transfer between processes. To reduce the likelihood of introducing storage bottlenecks, the data should be distributed evenly to the extent possible (assuming a homogeneous parallel computer). In Figure 6.7, we show examples of some commonly used distribution patterns for a matrix: one-dimensional partitionings of the matrix that distribute entire rows or columns among processes and a two dimensional partitioning, distributing matrix blocks. Each data distribution may have its own merits and drawbacks. For instance, a one-dimensional distribution of a matrix can take advantage of at most n processes for distribution of an $n \times n$ matrix, whereas a block distribution can distribute the matrix over as many as n^2 processes. Certain computational problems, however, may be more easily performed in parallel if an entire row or column of a matrix is available on one process (see section 6.4). Choosing a data distribution that reduces the amount of data transfer required during program execution will reduce the communication overhead and allow for a more efficient parallel implementation. This may entail using an owner-computes approach in which each process, to the extent possible, works only on the data stored locally. If a dynamic distribution scheme is employed, however, more extensive data retrieval from remote processes may be required.

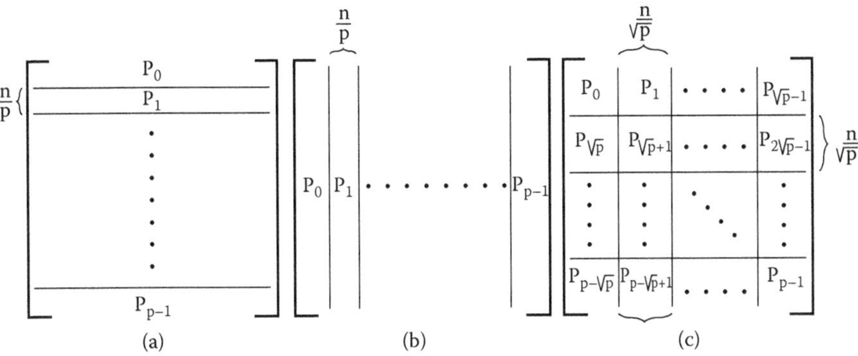

FIGURE 6.7
Distribution of an $n \times n$ matrix across p processes, $P_0 - P_{p-1}$, by (a) rows, (b) columns, and (c) square blocks. The one-dimensional distributions (a) and (b) can utilize at most n processes, whereas the two-dimensional block distribution (c) can distribute the matrix across up to n^2 processes.

Example 6.4 Data Distribution in the Two-Electron Integral Transformation

Consider data distribution in the two-electron integral transformation, which is a data-intensive computational step required in many quantum chemical methods. Chapter 9 presents a detailed discussion of parallel direct computation of MP2 energies, the dominant step of which is the two-electron integral transformation. Let us look at the integral transformation used in the P2 algorithm from chapter 9 and focus on how the data is distributed. The integral transformation must generate the set of two-electron integrals $(ia|jb)$ in the molecular orbital basis, distributed across processes in a manner that facilitates computation of the MP2 correlation energy from the expression

$$E_{\mathrm{MP2}}^{\mathrm{corr}} = \sum_{ijab} \frac{(ia|jb)[2(ia|jb) - (ib|ja)]}{\epsilon_i + \epsilon_j - \epsilon_a - \epsilon_b}. \tag{6.2}$$

This can be achieved by distributing the i, j pairs so that each process stores the $(ia|jb)$ integrals for a subset of the i, j pairs and all a, b pairs. The computation and transformation of the integrals proceed in separate steps as follows

$$(\mu\nu|\lambda\sigma) \rightarrow (i\nu|\lambda\sigma) \rightarrow (i\nu|j\sigma) \rightarrow (ia|j\sigma) \rightarrow (ia|jb) \tag{6.3}$$

and the distribution over i, j, therefore, cannot be performed until the half-transformed integrals $(i\nu|j\sigma)$ have been generated. The computation of the integrals in the atomic orbital basis, $(\mu\nu|\lambda\sigma)$, and the first quarter transformation, $(\mu\nu|\lambda\sigma) \rightarrow (i\nu|\lambda\sigma)$, involve nonuniform computational tasks, and a random (or, alternatively, dynamic) distribution of λ, σ is employed to achieve load balance. Only small batches of the

$(\mu\nu|\lambda\sigma)$ and $(i\nu|\lambda\sigma)$ integrals are generated at a time, so storage of these integrals is not a concern, but their distribution prevents the $(i\nu|j\sigma)$ integrals from being initially generated where they belong. A redistribution of these integrals is therefore required in the integral transformation (as shown in Figure 9.3). After this redistribution step, an owner-computes approach is possible, letting each process work independently on finishing the integral transformation and computing the contribution to the MP2 correlation energy for the locally stored integrals. The data distribution employed in this algorithm is uniform and allows for high parallel efficiency, requiring data exchange between processes only during the second quarter transformation.

6.3 Designing a Communication Scheme

Parallel program execution for anything but trivially parallel computational problems requires some exchange of data between processes. Parallel algorithm design therefore involves determining how this data exchange is to take place: it must be decided what data to exchange, when to do so, and what type of communication to use for this purpose. The communication requirement can be strongly dependent on the work and data distribution strategy, and this should be taken into account in the design phase. The exchange of large amounts of data between processes can take a significant amount of time and will adversely affect the parallel performance, and communication must therefore be employed judiciously. The use of collective communication operations, especially, can reduce the scalability of a parallel algorithm, and any communication step requiring synchronization of two or more processes tends to create idle time on some processes, lowering the efficiency. In general, and in particular when using collective communication, the communication to computation ratio should be kept as low as possible to minimize idle time on processes waiting for messages to be processed.

In the following we will discuss communication schemes employing collective and point-to-point operations and illustrate their use in parallel algorithms.

6.3.1 Using Collective Communication

The use of collective communication operations can simplify parallel programming. A complex exchange of data can be achieved with a single function call, and the programmer need not be concerned with the explicit sending and receiving of messages between individual processes. Although efficient implementations are readily available for many collective communication operations, their use must nonetheless be limited if high parallel performance is desired for large process counts. For collective operations involving replicated data, such as all-reduce and broadcast, the time required to do a collective communication operation increases with the number of processes involved and may become a bottleneck when the number of processes is large. Still, if only small amounts of data are involved, collective communication can be employed without serious performance penalty. Also, some collective operations have only a very small component that does not scale, and these operations may be suitable for use in a high-performance algorithm. An example of this is shown in section 10.3.1.

Example 6.5 Collective Communication Reduces Scalability
Collective communication operations can reduce the scalability of a parallel program by introducing a communication bottleneck. Consider, for example, an algorithm requiring n^2 floating point operations and using an all-to-all broadcast operation involving n^2/p pieces of data per process. The communication time for the all-to-all broadcast (using a recursive doubling algorithm) is $t_{\text{all-to-all}} = \alpha \log_2 p + \beta n^2(p-1)/p$ (from Eq. 3.4). The execution time on p processes can be expressed in terms of the machine parameters α, β, and γ as follows

$$t(p,n) = t_{\text{comp}}(p,n) + t_{\text{comm}}(p,n) = \gamma\frac{n^2}{p} + \alpha \log_2 p + \beta n^2 \frac{p-1}{p}. \quad (6.4)$$

The resulting speedup and efficiency are

$$S(p,n) = \frac{t(1,n)}{t(p,n)} = \frac{p}{1 + \frac{\alpha}{\gamma}\frac{p\log_2 p}{n^2} + \frac{\beta}{\gamma}(p-1)} \quad (6.5)$$

$$E(p,n) = \frac{S(p,n)}{p} = \frac{1}{1 + \frac{\alpha}{\gamma}\frac{p\log_2 p}{n^2} + \frac{\beta}{\gamma}(p-1)}. \quad (6.6)$$

The efficiency obtained is a decreasing function of the number of processes, and the algorithm is not strongly scalable. In fact, parallel algorithms employing collective communication are never strongly scalable. In this case, the algorithm is not guaranteed to be weakly scalable either; increasing the problem size cannot necessarily prevent a decrease in the efficiency as p increases because of the last term in the denominator in Eq. 6.6. Often, however, algorithms using collective communication are weakly scalable, and high efficiency can be obtained for larger process counts provided that the problem size is also large.

We will illustrate parallel matrix–vector multiplication algorithms using collective communication in section 6.4, and detailed examples and performance analyses of quantum chemistry algorithms employing collective communication operations can be found in sections 8.3, 9.3, and 10.3.

6.3.2 Using Point-to-Point Communication

We saw in section 6.3.1 that certain collective communication operations can introduce bottlenecks in parallel algorithms because the communication time is an increasing function of the number of processes. Point-to-point and one-sided communication offer a potential for higher performance for any type of communication involving enough data to make the communication time nonnegligible. The drawback of using these communication schemes is the increased complexity of parallel programming. The programmer must specify a detailed schedule for when and where to send and receive messages on each individual process while being careful to avoid pitfalls such as deadlock and race conditions. In section 9.4 one-sided communication is used to implement a high-performance second-order Møller–Plesset perturbation theory algorithm. In Example 6.6, we will illustrate the use of point-to-point message-passing, including a strategy for avoiding deadlock.

Example 6.6 Using a Systolic Loop Communication Pattern
Collective operations can be implemented in terms of point-to-point operations, and we will illustrate this use of point-to-point communication for a systolic loop algorithm: this algorithm exchanges data between neighboring processes in a systematic manner, and it could be used, for instance, to implement an all-to-all broadcast. In the systolic loop communication scheme illustrated in Figure 6.8, data is passed between neighboring processes in a virtual ring. An array of data, **a**, is assumed to be distributed by blocks across p processes, each process holding one block. The execution takes place in p stages that each consist of a computation step followed by a communication step. In the computation step every process performs some computation on the block of **a** currently residing on that process. In the subsequent communication step all processes exchange data with their neighbors: process P_i sends its

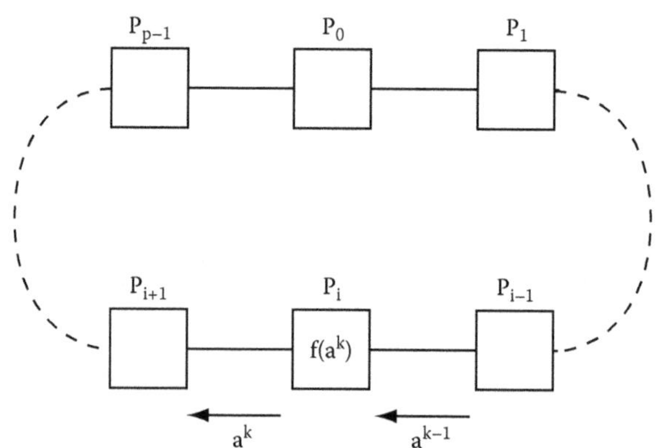

FIGURE 6.8
A systolic loop communication pattern using p processes, $P_0 - P_{p-1}$. Execution requires p steps, and in each step every process first performs computations on its local data and then exchanges data with its neighbors. The activity on process P_i in one step is shown: first, computation is performed on the locally residing block of data, $\mathbf{a}^k$, and $\mathbf{a}^k$ is then sent to the downstream neighbor while a new block, $\mathbf{a}^{k-1}$, is received from the upstream neighbor.

```
/* Determine which process to send to: */
int next_proc = (this_proc + 1) mod p

/* Determine which process to receive from: */
int previous_proc = (this_proc − 1 + p) mod p

int k = this_proc

For step = 0, step < p, step = step + 1
    Perform computation on local block (a^k) of a
    Post non-blocking receive of a^(k−1) from previous_proc
    Send a^k block to next_proc using blocking send
    Wait for completion of posted non-blocking receive
    k = (k + 1) mod p
End for
```

FIGURE 6.9
Outline of systolic loop communication scheme using point-to-point message-passing (cf. Figure 6.8). The number of processes and the process ID are designated p and this_proc, respectively. Non-blocking receive operations are employed to prevent deadlock.

current block of **a** to P_{i+1} while receiving a new block from P_{i-1} (process P_{p-1} sends data to P_0 to complete the loop). Because all processes try to exchange data with their neighbors at the same time, care must be exercised to avoid a deadlock situation. Deadlock can arise if blocking send and receive operations are used: if each process first posts a blocking receive (P_i receives from P_{i-1}) and then posts a blocking send (P_i sends to P_{i+1}), deadlock will result. A blocking receive operation posted by a process cannot be completed before the message has been received, but the message will not arrive because the sending process will be waiting for its own blocking receive operation to complete. Therefore, the systolic loop communication scheme must be implemented using non-blocking receives (or sends): a non-blocking receive must be issued, followed by a blocking send, and, finally, a wait for the receive operation to complete, as shown in Figure 6.9. By using a non-blocking receive, the subsequent blocking send operation can be initiated before the receive is complete, preventing deadlock. The wait operation posted after the send waits for the receive to complete before proceeding to ensure that data has been received before it is used.

6.4 Design Example: Matrix-Vector Multiplication

To illustrate some of the parallel program design principles discussed in this chapter, let us consider the design of a parallel algorithm to perform a dense matrix–vector multiplication $\mathbf{Ab} = \mathbf{c}$, where $\mathbf{A}$ is an $n \times n$ matrix and $\mathbf{b}$ and $\mathbf{c}$ are n-dimensional vectors. To be able to take full advantage of the parallel resources, we will require that the matrix **A** is distributed and that the number of processes that can be utilized is at least equal to the dimension n. Additionally, we will assume that the final distribution of the **c** vector is required to be the same as the initial distribution of **b**. For a dense matrix–vector multiplication, the computational tasks are uniform, so a static task distribution will be adequate. We will first look at a parallel algorithm using a row-distribution for the matrix, and we will then explore another algorithm that distributes the matrix by blocks.

6.4.1 Using a Row-Distributed Matrix

Let us first try a simple parallelization scheme in which **A** is distributed by rows across p processes, $P_0 - P_{p-1}$, and the vector **b** is replicated as illustrated in Figure 6.10 (assuming n is a multiple of p). During the multiplication, the product vector **c** will be distributed, but upon completion of the multiplication, we want the **c** vector to be replicated as well. The elements, c_i, of **c** are computed as the product of an entire row of **A** and the **b** vector

$$c_i = \sum_j A_{ij} b_j. \tag{6.7}$$

Because the process P_i holds the rows $i \times n/p$ through $(i+1) \times n/p - 1$ of **A**, P_i can compute the corresponding elements of **c**, $c_{i\times n/p} - c_{(i+1)\times n/p-1}$, using only locally stored data. Each process first independently computes the elements of **c** corresponding to the locally stored rows of **A**. When this step is complete, a single collective communication step is required to put a copy of the entire **c** vector on all processes. This can be accomplished with an all-to-all broadcast operation in which each process sends its elements of **c** to all other processes. An outline of this algorithm can be found in chapter 5 (Figure 5.4) where matrix–vector multiplication is discussed in the context of performance modeling.

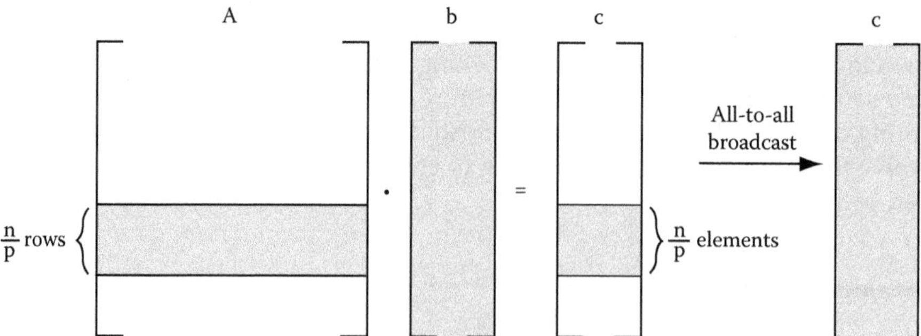

FIGURE 6.10
Data residing on process P_i in a parallel matrix–vector multiplication $\mathbf{Ab} = \mathbf{c}$, where **A** is a row-distributed $n \times n$ matrix, and **b** and **c** are replicated vectors of length n. The process count is p, and P_i holds the rows of **A** numbered in/p through $(i+1)n/p-1$ and computes the corresponding elements of **c**. A final all-to-all broadcast puts the entire **c** vector on all processes.

Let us briefly consider the performance characteristics of this algorithm. The degree of parallelism equals n because at most n processes can be used to distribute the rows of **A**. The memory requirement per process is $n^2/p + 2n$ (requiring storage of n/p rows of **A** and the entire **b** and **c** vectors). The communication overhead is the time required for an all-to-all broadcast involving p processes and a data length of n/p. Using a recursive doubling algorithm and assuming that p is a power of two, the communication time is $t_{\text{comm}} = \alpha \log_2 p + \beta n(p-1)/p$ (Eq. 3.4). The total number of floating point operations required to perform the multiplication is n^2 (counting a combined multiply and add as one operation), and the execution time can therefore be expressed as

$$t(p,n) = t_{\text{comp}}(p,n) + t_{\text{comm}}(p,n) = \gamma \frac{n^2}{p} + \alpha \log_2 p + \beta n \frac{p-1}{p} \tag{6.8}$$

using the machine parameters α, β, and γ. This yields the following speedup and efficiency

$$S(p,n) = \frac{\gamma n^2 p}{\gamma n^2 + \alpha p \log_2 p + \beta n(p-1)} \tag{6.9}$$

$$E(p,n) = \frac{S(p,n)}{p} = \frac{1}{1 + \frac{\alpha}{\gamma}\frac{p \log_2 p}{n^2} + \frac{\beta}{\gamma}\frac{p-1}{n}}. \tag{6.10}$$

From Eq. 6.10 it follows that the dimension n must grow at the same rate as p to maintain a constant efficiency as the number of processes increases. If n increases at the same rate as p, however, the memory requirement per process $(n^2/p + 2n)$ will increase with the number of processes. Thus, a k-fold increase in p, with a concomitant increase in n to keep the efficiency constant, will lead to a k-fold increase in the memory required per process, creating a potential memory bottleneck. Measured performance data for a parallel matrix–vector multiplication algorithm using a row-distributed matrix are presented in section 5.3.2.

6.4.2 Using a Block-Distributed Matrix

Let us try to modify the matrix–vector multiplication algorithm from section 6.4.1 to improve the scalability. The poor scalability was a result of the relatively large communication overhead incurred by using a row distribution for the matrix **A**. When **A** is distributed by rows, all elements of the **b** or **c** vector must visit (or be stored by) each process during the computation: if **b** and **c** are replicated, no data exchange is required for **b**, but an all-to-all broadcast is required to replicate **c** at the end of the computation; if both vectors are distributed, no communication is required for **c** but all elements of **b** must visit all processes during the execution.

The first requirement for the new algorithm therefore is to distribute **A** differently. A column distribution will not reduce the communication requirement, so we need to distribute **A** by blocks. We will employ the data distribution illustrated in Figure 6.11, assuming that n is a multiple of p, and that p is a square number. The p processes are numbered $\mathrm{P}_{i,j}$ using indices i and j that both run from 0 to $\sqrt{p}-1$, and in Figure 6.11 the processes are shown on a two-dimensional grid with the process $\mathrm{P}_{i,j}$ positioned in row i and column j of each grid. The matrix $\mathbf{A}$ is divided into p square blocks $\mathbf{A}^{i,j}$, each with dimensions $n/\sqrt{p} \times n/\sqrt{p}$, and process $\mathrm{P}_{i,j}$ holds the block $\mathbf{A}^{i,j}$. The vector $\mathbf{b}$ is divided into $\sqrt{p}$ blocks of size $n/\sqrt{p}$ and distributed across the first row of processes, $P_{0,j}, 0 \leq j < \sqrt{p}$. The final distribution of $\mathbf{c}$ is identical to the initial distribution of $\mathbf{b}$.

Using these data distributions, the matrix–vector multiplication proceeds as shown in Figures 6.11 and 6.12. Initially, every process in the first row performs a one-to-all broadcast of its block of $\mathbf{b}$ to all other processes in the same column. Every process $\mathrm{P}_{i,j}$ now holds the blocks $\mathbf{A}^{i,j}$ and $\mathbf{b}^{j}$, and $\mathrm{P}_{i,j}$ then computes the product $\mathbf{A}^{i,j}\mathbf{b}^{j}$ of these locally stored blocks, producing the contribution $\mathbf{c}^{i(j)}$to the block $\mathbf{c}^{i}$. The processes in row i now hold all the contributions to the block $\mathbf{c}^{i}$, and for each row these contributions are added in an all-to-one reduce operation, putting the resulting completed block $\mathbf{c}^{i}$ on the process $\mathrm{P}_{i,0}$. Finally, after a point-to-point send of the block $\mathbf{c}^{i}$ from $\mathrm{P}_{i,0}$ to $\mathrm{P}_{0,i}$, the matrix–vector multiplication is complete, with each of the first $\sqrt{p}$ processes holding a block of $\mathbf{c}$.

The communication overhead for this algorithm is the cost of performing a broadcast and a reduce operation that both involve $n/\sqrt{p}$ elements and $\sqrt{p}$ processes (ignoring the small cost of the final point-to-point send putting the block $\mathbf{c}^{i}$ on P_i). Ignoring the computation needed by the reduction, the cost for these two operations is the same, and the total communication time is $t_{\mathrm{comm}} = 2(\alpha + \beta n/\sqrt{p}) \log_2 \sqrt{p} = (\alpha + \beta n/\sqrt{p}) \log_2 p$, using a binomial tree algorithm (Eq. 3.1) and assuming that p is a power of two. Note that the communication requirement for the block distribution will be minimized by dividing the matrix into square blocks, as done here. We then get the following expressions for the execution time, speedup, and efficiency

$$t(p,n) = \gamma \frac{n^2}{p} + \log_2 p(\alpha + \beta n/\sqrt{p}) \tag{6.11}$$

$$S(p,n) = \frac{\gamma n^2 p}{\gamma n^2 + \alpha p \log_2 p + \beta n \sqrt{p} \log_2 p} \tag{6.12}$$

$$E(p,n) = \frac{1}{1 + \frac{\alpha}{\gamma}\frac{p \log_2 p}{n^2} + \frac{\beta}{\gamma}\frac{\sqrt{p}\log_2 p}{n}}. \tag{6.13}$$

For a constant problem size, the efficiency for this algorithm is a decreasing function of p, so the algorithm is not strongly scalable. To maintain the efficiency, however, the dimension n only needs to grow proportionally to $\sqrt{p}\log_2 p$, whereas n was required to increase proportionally to p in the algorithm using a row distribution for $\mathbf{A}$. Because $\sqrt{p}\log_2 p$ grows significantly more slowly than p, this represents an improvement in scalability. From Eqs. 6.10 and 6.13 it follows that for a given dimension n, the efficiency obtained using the block distribution is higher than that obtained using the

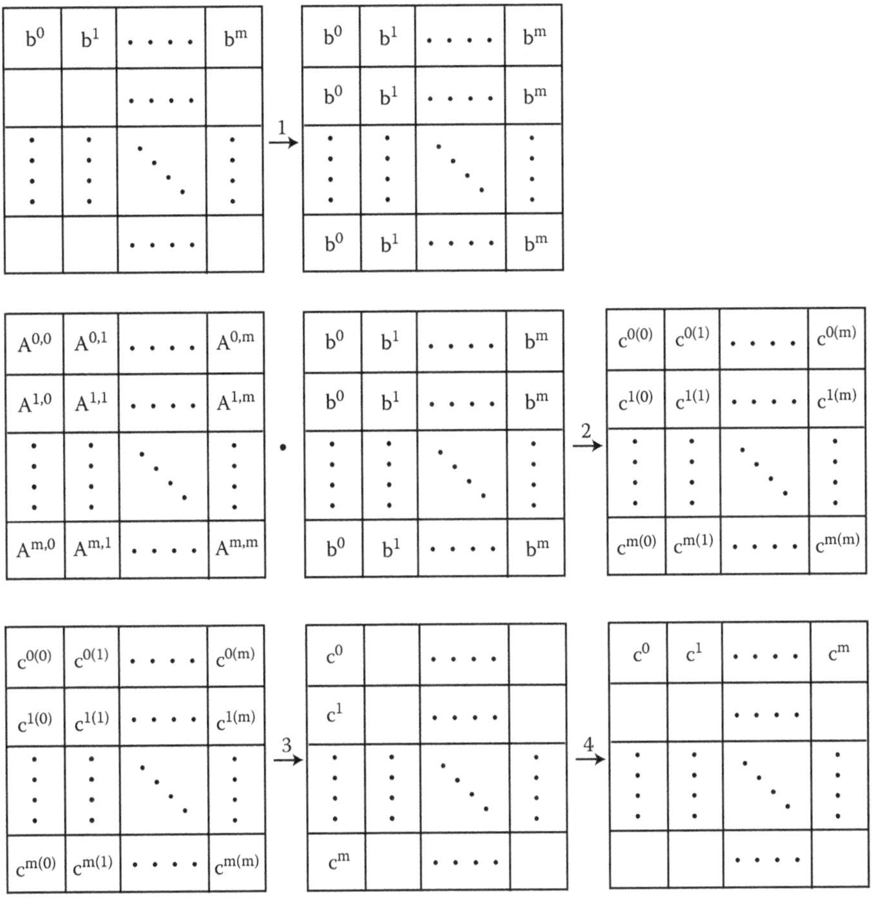

FIGURE 6.11
Parallel matrix–vector multiplication $\mathbf{Ab} = \mathbf{c}$ on p processes using fully distributed data. The processes are shown on a grid with process $P_{i,j}$ in row i and column j $(0 \le i < \sqrt{p}, 0 \le j < \sqrt{p})$. The steps indicated with arrows are: (1) an initial broadcast within each column putting a copy of $\mathbf{b}^j$ on every process in column j; (2) computation of the contribution $\mathbf{c}^{i(j)} = \mathbf{A}^{i,j}\mathbf{b}^j$ to the block $\mathbf{c}^i$ on $P_{i,j}$; (3) summation of $\mathbf{c}^{i(j)}$ across each row, putting the complete block $\mathbf{c}^i$ on $P_{i,0}$; and (4) redistribution of $\mathbf{c}$ ($P_{i,0}$ sends $\mathbf{c}^i$ to $P_{0,i}$) to match the initial distribution of $\mathbf{b}$. The index m represents the value $\sqrt{p} - 1$. $\mathbf{A}$ is distributed by square blocks, and $P_{i,j}$ holds the block $\mathbf{A}^{i,j}$. The vectors $\mathbf{b}$ and $\mathbf{c}$ are divided into uniform blocks, and $P_{0,j}$ holds the block $\mathbf{b}^j$ and (at the end) $\mathbf{c}^j$.

row distribution of **A** for larger process counts ($p > 21$). For smaller numbers of processes, however, the efficiency obtained using a row distribution is slightly higher. In addition to higher scalability, the algorithm using the block-distributed **A** also has a much higher degree of parallelism, being able to use as many as n^2 processes. Furthermore, if n grows at the rate required to keep up the efficiency, $\sqrt{p}\log_2 p$, the memory requirement per process does not increase as fast as for the first algorithm, reducing the likelihood of a memory bottleneck.

j = this_proc mod $\sqrt{p}$ (column index for $P_{i,j}$)
i = (this_proc − j)/$\sqrt{p}$ (row index for $P_{i,j}$)

If i eq 0: Broadcast block b^j to $P_{k,j}$ (for $0 < k < \sqrt{p}$)

Compute local contribution to block c^i: $c^{i(j)} = A^{i,j}\, b^j$

Do a reduce operation adding $c^{i(j)}$ across each row to collect full c^i on $P_{i,0}$

If j eq 0: Send c^i to $P_{0,i}$ (final **c** distribution: c^i on $P_{0,i}$)

FIGURE 6.12
Outline of the algorithm for parallel matrix–vector multiplication $\mathbf{Ab} = \mathbf{c}$ with block-distributed matrix $\mathbf{A}$ discussed in the text. The number of processes is designated p, and this_proc is the process ID. The employed data distribution is shown in Figure 6.11; $\mathbf{b}^j$ and $\mathbf{c}^i$ represent blocks of length $n/\sqrt{p}$ of the **b** and **c** vectors, respectively, and $\mathbf{A}^{i,j}$ is the block of **A** stored by process $P_{i,j}$.

6.5 Summary of Key Points of Parallel Program Design

In the previous sections of this chapter, we have discussed the design of parallel programs, including the distribution of work and data as well as the design of a communication scheme. The design phase also should include at least a preliminary performance model for the proposed parallel algorithm so that it can be ascertained whether the program is likely to meet the various desired performance criteria. Performance modeling for parallel programs is discussed in detail in chapter 5. If a preliminary performance model for a proposed parallel algorithm reveals serious flaws, one may revisit the previous design steps and possibly reach a compromise between different design goals to achieve a satisfactory algorithm. Below we give an overview of the steps involved in parallel program design, summarizing some of the key points discussed in this chapter and also including some considerations regarding gauging the parallel efficiency of an algorithm.

- Distribution of Work
  - Consider desired degree of parallelism
  - Partition computational problem into smaller, preferably uniform, tasks that can be performed concurrently by individual processes
  - Decide whether to distribute work statically or dynamically
    - Are task sizes known in advance? Are they uniform?
  - If using static task distribution
    - Select a distribution scheme, for example, round-robin or recursive
    - If tasks are nonuniform, randomization of tasks before distribution may improve load balance

- If using dynamic task distribution
  - Select a distribution scheme, for example, manager–worker or decentralized
  - If tasks are nonuniform, initial sorting of tasks according to size and distribution from largest to smallest usually improves load balance

- Distribution of Data
  - Decide what data must be distributed
    - Large data structures should be distributed to avoid memory bottlenecks
    - Smaller data structures can possibly be replicated to reduce communication overhead and the need for synchronization and to simplify parallelization
  - Decide how to distribute data
    - Consider whether the task distribution suggests a particular data distribution
    - Consider distributing data in a way that minimizes the communication overhead
    - Can data be distributed so processes do computation only on local data?
- Designing a Communication Scheme
  - Decide whether to use collective, point-to-point, or one-sided communication
    - How much data exchange must take place?
    - Can collective communication be used without introducing a communication bottleneck?
    - Collective communication simplifies programming but tends to reduce scalability, particularly when replicated data is involved
    - Point-to-point or one-sided communication may improve performance but complicates implementation
  - If using collective communication
    - Try to minimize the amount of data to be communicated in collective operations
    - Collective operations require synchronization of processes; carefully consider where to put them
  - If using point-to-point communication
    - Decide whether to use blocking or non-blocking operations
    - Determine message sizes, possibly coalescing many small messages into fewer larger ones
    - Carefully check to avoid deadlock and race conditions
    - Check for proper termination detection
    - Try to overlap communication and computation

  - If using manager–worker scheme, consider if a bottleneck is likely on manager
- Performance Modeling
  - Work out performance model for the parallel algorithm (this can be just a rough model, but it should try to capture the essential performance features)
  - Is parallel performance adequate? Efficiency? Scaling?
  - Is load imbalance likely to be an issue? Is it accounted for in the model?
  - Try to identify potential bottlenecks (serial code, replicated arrays, collective communication steps)
  - Is the computation-to-communication ratio acceptable?
  - Consider tradeoffs; for example, partial data replication may reduce communication, and sacrificing scalar speed may offer potential for improved parallel implementation

6.6 Further Reading

For a discussion of potential benefits of parallelism and how to decide whether parallelism might be worthwhile for a given application, see Pancake.[5] An approach to methodical design of parallel algorithms, involving four steps designated partitioning, communication, agglomeration, and mapping, has been introduced by Foster.[1] As mentioned in section 6.1.2.2, termination detection can be a non-trivial task when using a decentralized task distribution. For an overview of termination detection algorithms, see, for example, Tel.[6]

References

1. Foster, I. Designing and building parallel programs [online]. http://www-unix.mcs.anl.gov/dbpp.
2. Grama, A., A. Gupta, G. Karypis, and V. Kumar. *Introduction to Parallel Computing*, 2nd edition. Harlow, England: Addison-Wesley, 2003.
3. Quinn, M. J. *Parallel Programming in C with MPI and OpenMP*. New York: McGraw-Hill, 2003.
4. Hoare, C. A. R. Quicksort. *Comp. J.* 5:10–15, 1962.
5. Pancake, C. M. Is parallelism for you? *IEEE Comp. Sci. Eng.* 3:18–37, 1996.
6. Tel, G. *Introduction to Distributed Algorithms*, chapter 8. Cambridge: Cambridge University Press, 2000.

Part II

Applications of Parallel Programming in Quantum Chemistry

7

Two-Electron Integral Evaluation

The two-electron integrals, also known as electron repulsion integrals, are ubiquitous in quantum chemistry, appearing in the Hartree–Fock method and any correlated method based upon it as well as in density functional methods. The two-electron integral computation is a time-consuming step, and an efficient integral program is an important part of a quantum chemistry code. When designing parallel algorithms for computation of the two-electron integrals, assuming that an efficient scalar code exists, issues of particular importance are load balancing of the nonuniform computational tasks involved as well as utilization of the permutational symmetry of the integrals. In this chapter we will first discuss a few basic issues pertaining to the computation of the two-electron integrals, and we will then examine different parallel implementations of the integral computation, using either a simple static load balancing scheme or employing a manager–worker model for dynamically distributing the work.

7.1 Basics of Integral Computation

The two-electron integrals arise in the evaluation of matrix elements between Slater determinants of a two-electron operator of the form

$$\frac{1}{|\mathbf{r}_i - \mathbf{r}_j|} = \frac{1}{r_{ij}} \tag{7.1}$$

which represents electron–electron repulsion. Using the so-called chemist's, or Mulliken, notation, the two-electron integral denoted $(\chi_\mu \chi_\nu | \chi_\rho \chi_\lambda)$ is defined as

$$(\chi_\mu \chi_\nu | \chi_\rho \chi_\lambda) = \int \phi_\mu^*(\mathbf{r}_1)\sigma_\mu^*(\omega_1)\phi_\nu(\mathbf{r}_1)\sigma_\nu(\omega_1) r_{12}^{-1} \phi_\rho^*(\mathbf{r}_2) \times \sigma_\rho^*(\omega_2)\phi_\lambda(\mathbf{r}_2)\sigma_\lambda(\omega_2) d\mathbf{r}_1 d\omega_1 d\mathbf{r}_2 d\omega_2 \tag{7.2}$$

where $\mathbf{r}_i$ and ω_i represent the spatial and spin coordinates, respectively, of electron i, $\chi = \phi(\mathbf{r})\sigma(\omega)$ is a spin orbital, ϕ denotes a spatial orbital, and σ designates a spin function (either α or β). Integrating over the spin of the electrons and using real orbitals, the two-electron integral takes on the following form in the spatial orbital basis

$$(\phi_\mu \phi_\nu | \phi_\rho \phi_\lambda) = \int \phi_\mu(\mathbf{r}_1)\phi_\nu(\mathbf{r}_1) r_{12}^{-1} \phi_\rho(\mathbf{r}_2)\phi_\lambda(\mathbf{r}_2) d\mathbf{r}_1 d\mathbf{r}_2. \tag{7.3}$$

We will use the shorthand notation $(\mu\nu|\rho\lambda)$ to represent the integral $(\phi_\mu\phi_\nu|\phi_\rho\phi_\lambda)$ from Eq. 7.3, where μ, ν, ρ, and λ label real, spatial atomic orbitals.

The number of two-electron integrals in the atomic orbital basis equals n^4, where n is the number of basis functions. The two-electron integrals possess permutational symmetry, however, and the integral $(\mu\nu|\rho\lambda)$ is invariant to the following exchanges of indices: $\mu \leftrightarrow \nu$, $\rho \leftrightarrow \lambda$, and $\mu\nu \leftrightarrow \rho\lambda$. This yields an eight-fold permutational symmetry in the two-electron integrals, and the number of unique integrals, therefore, is only $\approx n^4/8$ (the exact number of unique integrals is $\frac{1}{4}n(n+1)[\frac{1}{2}n(n+1)+1]$). For computational efficiency, related basis functions (atomic orbitals) are grouped into shells, and the two-electron integrals are computed for one shell quartet at a time. Thus, if M, N, R, and S represent shells of atomic orbitals, all integrals $(\mu\nu|\rho\lambda)$ with $\mu \in M$, $\nu \in N$, $\rho \in R$, and $\lambda \in S$ for a fixed M, N, R, S shell quartet are computed together. All basis functions constructed from the same set of primitive Gaussian functions are included in a shell, and a shell may represent one angular momentum, for instance, an f shell, or a mixture of angular momenta, such as an sp shell.

In practice, integral screening is employed before the integrals are computed. The screening is performed using the Schwarz inequality

$$|(\mu\nu|\rho\lambda)| \leq (\mu\nu|\mu\nu)^{1/2}(\rho\lambda|\rho\lambda)^{1/2}. \tag{7.4}$$

If the integrals $(\mu\nu|\rho\lambda)$ are found to be below a certain threshold for all μ, ν, ρ, λ in the M, N, R, S quartet, the entire shell quartet can be neglected. This screening can significantly reduce the number of integrals that must be evaluated, and for extended systems the number of nonnegligible integrals approaches $O(n^2)$.[1] Integral evaluation, however, is still a computationally expensive step, often the dominant step in conventional Hartree–Fock and density functional methods, and utilization of the permutational symmetry of the integrals is important to further reduce the computational work.

The computational expense of evaluating the integrals in a shell quartet is strongly dependent on the angular momenta represented in the quartet. The computation time is larger for shells with basis functions of high angular momentum because these shells contain more basis functions and because the evaluation of the integral for each of these functions generally is more costly. We note that for a shell with a total angular momentum of l, the number of basis functions (using spherical harmonics) is $2l+1$, and each of these functions is a product of a polynomial of degree l and a linear combination of

Gaussian functions. For instance, evaluation of a shell quartet of f functions, $(ff|ff)$, may take more than a thousand times longer than that of a pure s quartet, $(ss|ss)$. The exact time required to compute a shell quartet of integrals, however, is usually not known in advance. When designing parallel algorithms for two-electron integral computation, one is therefore faced with the challenge of creating an even distribution over a number of processes of tasks whose sizes vary widely but are not known exactly.

In the remainder of this chapter we will discuss different ways to parallelize the computation of the two-electron integrals with focus on how to achieve load balance. Two-electron integral computation in a quantum chemistry code is often performed in the context of a procedure such as Fock matrix formation or two-electron integral transformation, but we here consider the integral computation separately to specifically concentrate on strategies for distributing the work. Examples of parallel two-electron integral computation in quantum chemical methods are discussed in chapters 8, 9, and 10.

7.2 Parallel Implementation Using Static Load Balancing

The most straightforward way to distribute the work in the two-electron integral computation is to employ a static load balancing scheme that distributes the tasks in a round-robin fashion. Because all the integrals belonging to a shell quartet are computed together, the smallest possible tasks that can be distributed are individual shell quartets. The distribution of individual shell quartets is advantageous in terms of achieving load balance because, given the nonuniformity of the computational tasks, breaking the tasks into the smallest possible chunks will yield a more even distribution. In parallel electronic structure programs, however, distribution of individual shell quartets is not always feasible. For instance, if the two-electron integrals will subsequently be transformed from the atomic to the molecular orbital basis, the memory and communication requirements can be reduced significantly by distributing shell pairs instead of shell quartets. We will here analyze the performance of two parallel algorithms for two-electron integral computation, one algorithm distributing individual shell quartets and the other algorithm distributing shell pairs. We note that, while distribution of just one shell index is a very straightforward way to parallelize the two-electron integral computation, this distribution yields only a small number of tasks, which are irregular, and load imbalance will cause rapid performance degradation as the number of processes increases.

7.2.1 Parallel Algorithms Distributing Shell Quartets and Pairs

Two parallel algorithms for two-electron integral computation using static load balancing are outlined in Figure 7.1. The algorithms, designated (a) and (b), both employ a static work distribution scheme with a round-robin

```
i = 0
For M = 1, n_shell
    For N = 1, M
        For R = 1, M
            If R eq M: S_max = N
            Else: S_max = R
            For S = 1, S_max
                If i mod p eq this_proc
                    Compute (MN|RS)
                Endif
                i = i + 1
            End for
        End for
    End for
End for
```

(a)

```
i = 0
For M = 1, n_shell
    For N = 1, M
        If i mod p eq this_proc
            For R = 1, M
                If R eq M: S_max = N
                Else: S_max = R
                For S = 1, S_max
                    Compute (MN|RS)
                End for
            End for
        Endif
        i = i + 1
    End for
End for
```

(b)

FIGURE 7.1
Outline of two parallel algorithms for two-electron integral computation using static load balancing. Algorithm (a) distributes shell quartets $MNRS$, and algorithm (b) distributes shell pairs MN. Both algorithms utilize the full integral permutation symmetry. The number of shells and the process count are denoted n_{shell} and p, respectively, and this_proc ($0 \leq$ this_proc $< p$) represents the process identifier.

allocation of tasks, require no communication, and utilize the full eight-fold permutational symmetry of the integrals. The algorithms differ in the sizes of the computational tasks that are distributed, however: algorithm (a) distributes individual shell quartets $MRNS$, whereas algorithm (b) uses a distribution of shell pairs MN. In algorithm (b), once an MN shell pair has been assigned to a process, this process computes all the integrals $(MN|\overline{RS})$ for that MN pair, where $\overline{RS}$ represents all included RS shell pairs. The number of tasks to be distributed in algorithm (a) is the number of shell quartets, n_{MNRS}, and in algorithm (b) the number of tasks equals the number of MN shell pairs, n_{MN}. Expressed in terms of the number of shells in the basis set, n_{shell}, the number of tasks in algorithms (a) and (b) are $\approx n_{\text{shell}}^4$ and $\approx n_{\text{shell}}^2$, respectively (ignoring constant factors arising from the utilization of permutational symmetry), and the degree of parallelism, thus, is much higher for algorithm (a). When Schwarz screening (Eq. 7.4) is employed, the number of tasks in (a) and (b) become $\propto n_{\text{shell}}^2$ and $\propto n_{\text{shell}}$, respectively, for large molecules. In the next section we will present a detailed performance analysis of the two algorithms.

7.2.2 Performance Analysis

To analyze the performance of algorithms (a) and (b), we will first obtain expressions for the parallel efficiencies. For these algorithms, which are fully parallelized and involve no communication, load imbalance is the only factor that may contribute significantly to lowering the parallel efficiency, and to predict the parallel performance, we must be able to estimate this load imbalance. Additionally, to make quantitative predictions for the efficiency, it is necessary to collect some statistics for the computational times required for evaluation of the integrals in a shell quartet.

Let us first consider how to model the load imbalance. We know that the times required to compute the two-electron integrals for individual shell quartets $(MN|RS)$ vary, and we will consider these times as a distribution with mean μ_q and standard deviation σ_q. Likewise, the computation times for all the integrals corresponding to fixed MN shell pairs, $(MN|\overline{RS})$, form a distribution with mean μ_p and standard deviation σ_p. For algorithm (a), a round-robin distribution of shell quartets using p processes will yield about n_{MNRS}/p shell quartets to be handled by each process. We will assume that the processing time for each of these tasks is selected from a random distribution described by the appropriate μ and σ. This is a good approximation if the number of tasks is much greater than p, and, in turn, $p \gg 1$. It then follows from basic probability theory that the total process execution times in algorithm (a) will form a distribution with mean $\mu_a = \mu_q \times n_{MNRS}/p$ and standard deviation $\sigma_a = \sigma_q \times \sqrt{n_{MNRS}/p}$. Analogously, for algorithm (b) random distribution of n_{MN} tasks (shell pairs) over p processes yields a distribution of process execution times with mean $\mu_b = \mu_p \times n_{MN}/p$ and standard deviation $\sigma_b = \sigma_p \times \sqrt{n_{MN}/p}$.

We will now use these distributions to express the total execution time, that is, the maximum execution time for any one process, for algorithms (a) and (b). This time can be approximated by a sum of the average execution time and a load imbalance term proportional to the standard deviation as follows

$$
\begin{aligned}
t_a(p, n_{MNRS}) &= \mu_a(p, n_{MNRS}) + k(p)\sigma_a(p, n_{MNRS}) \\
&= \mu_q \frac{n_{MNRS}}{p} + k(p)\sigma_q \sqrt{\frac{n_{MNRS}}{p}}
\end{aligned}
\tag{7.5}
$$

$$
\begin{aligned}
t_b(p, n_{MN}) &= \mu_b(p, n_{MN}) + k(p)\sigma_b(p, n_{MN}) \\
&= \mu_p \frac{n_{MN}}{p} + k(p)\sigma_p \sqrt{\frac{n_{MN}}{p}}.
\end{aligned}
\tag{7.6}
$$

We have here expressed the load imbalance as the standard deviation σ times a factor $k(p)$, which is a function of the number of processes, and we will determine the functional form for $k(p)$ in section 7.2.2.1. From Eqs. 7.5 and

7.6, we get the following expressions for the efficiencies

$$
\begin{aligned}
E_a(p, n_{MNRS}) &= \frac{t_a(1, n_{MNRS})}{p t_a(p, n_{MNRS})} \\
&= \frac{1 + k(1)(\sigma_q/\mu_q)/\sqrt{n_{MNRS}}}{1 + k(p)(\sigma_q/\mu_q)\sqrt{p/n_{MNRS}}}
\end{aligned}
\tag{7.7}
$$

$$
\begin{aligned}
E_b(p, n_{MN}) &= \frac{t_b(1, n_{MN})}{p t_b(p, n_{MN})} \\
&= \frac{1 + k(1)(\sigma_p/\mu_p)/\sqrt{n_{MN}}}{1 + k(p)(\sigma_p/\mu_p)\sqrt{p/n_{MN}}}.
\end{aligned}
\tag{7.8}
$$

By means of Eqs. 7.7 and 7.8 the scalability of the algorithms can be investigated. We first note that $k(p)$ is a slowly varying function of p and can be assumed to be constant for the purpose of scalability analysis (see section 7.2.2.1). The algorithms are weakly scalable because the efficiency can be maintained as the number of processes increases provided that the problem size increases as well. For algorithm (a) to maintain a nearly constant efficiency as p increases, the number of shell quartets, n_{MNRS}, must grow at the same rate as p, while for algorithm (b), maintaining a constant efficiency requires the number of shell pairs, n_{MN}, to increase proportionally to p. With n_{shell} shells in the basis set, n_{MNRS} and n_{MN} will be proportional to n^4_{shell} and n^2_{shell}, respectively. Thus, in algorithm (a), n_{shell} needs to increase only as $p^{1/4}$ to keep the efficiency from decreasing, whereas in algorithm (b) n_{shell} must increase as $p^{1/2}$. Likewise, when integral screening is employed, n_{shell} must increase proportionally to $p^{1/2}$ and p for (a) and (b), respectively, to maintain a constant efficiency. This difference in scalability for the two algorithms has important consequences for the parallel performance as the number of processes gets large.

To employ the performance models in Eqs. 7.7 and 7.8 for quantitive predictions of the parallel performance, we need an expression for $k(p)$ as well as the means and standard deviations associated with the integral computation. We will discuss below how to obtain these parameters, and we will illustrate both the predicted and the actual, measured performance for the two algorithms.

7.2.2.1 Determination of the Load Imbalance Factor $k(p)$

To obtain an expression for $k(p)$ we will assume that the process execution times for both algorithms (a) and (b) form a normal distribution, which is a reasonable assumption (according to the Central Limit Theorem from probability theory) when there is a large number of tasks per process. Assuming a normal distribution with mean μ and standard deviation σ, the probability of a process execution time being below $\mu + k\sigma$ can be computed as $\frac{1}{2} + \frac{1}{2}\text{erf}(k/\sqrt{2})$, where erf denotes the error function. If there are p processes, the probability that all process execution times are below $\mu + k\sigma$ is then given as $[\frac{1}{2} + \frac{1}{2}\text{erf}(k/\sqrt{2})]^p$. We need Eqs. 7.5 and 7.6 to be fairly accurate estimates for the maximum execution time, and we must therefore choose k such that the probability that all process execution times are below $\mu + k\sigma$ is nearly unity. Let this probability be represented by T. We then need to solve the following equation for k

$$\left[\frac{1}{2} + \frac{1}{2} \times \text{erf}(k/\sqrt{2})\right]^p = T \tag{7.9}$$

which yields

$$k = \sqrt{2} \times \text{erf}^{-1}(2T^{1/p} - 1). \tag{7.10}$$

In our performance models we will use a value of $T = 0.99$, that is, we will require that the probability that no process execution time exceeds that of the model equals 99%. Some care must be exercised in choosing an appropriate T value. A value too close to unity will cause Eqs. 7.5 and 7.6 to predict unrealistically high total execution times because there is no upper limit for the normal distribution. If T is too small, on the other hand, Eqs. 7.5 and 7.6 are likely to underestimate the total execution time. For simplicity, we will use an approximate form for the inverse error function, and for $T = 0.99$ we find the following expression to be a very good approximation to Eq. 7.10

$$k(p) \approx \sqrt{1.9 \times \ln p + 5.3}. \tag{7.11}$$

We will use this functional form for $k(p)$ in the performance models for algorithms (a) and (b) below. The term $\sqrt{1.9 \times \ln p + 5.3}$ increases very slowly with p and is nearly constant over a wide range of process counts. Note that the statistical approach employed in modeling the load imbalance entails using a $k(p)$ function that does not tend to 0 as p approaches 1, and the model breaks down for very small process counts. However, if Eqs. 7.5 and 7.6 were employed simply to predict the single-process timings, a value of 0 should be used for $k(1)$.

7.2.2.2 Determination of μ and σ for Integral Computation

To compute the total execution times and efficiencies from Eqs. 7.5–7.8, the means, μ_q and μ_p, and standard deviations, σ_q and σ_p, are required. In Table 7.1 we show measured values for these quantities for a number of commonly used basis sets, ranging from the minimal set STO-3G to the large correlation-consistent quadruple-ζ set cc-pVQZ. The values reported here were determined on a single processor of a Linux cluster[2] for the ethane molecule using the default two-electron integrals program in the MPQC program suite.[3] The values may show considerable variation with the employed integrals program and hardware, which should therefore be chosen to be representative of what will be employed in the parallel application. For predicting the parallel efficiency, only the ratio σ/μ is required. For individual shell quartets $MNRS$, the ratio σ_q/μ_q ranges from about 0.5 for a minimal basis set to 2–4 for large basis sets. When considering all integrals $(MN|\overline{RS})$ together, the ratio of the standard deviation to the mean, σ_p/μ_p shows less variation and assumes values in the range 0.6–1.5 for the basis set studied.

TABLE 7.1

The average time (μ) and the ratio of the standard deviation to the average time (σ/μ) for computation of shell quartets of two-electron integrals. Results were obtained for the ethane molecule on a single processor of a Linux cluster.[2] The number of basis functions and shells in the basis set are denoted n and n_{shell}, respectively. The subscript q refers to individual shell quartets of integrals, $(MN|RS)$, whereas the subscript p refers to sets of integrals $(MN|\overline{RS})$ for one MN pair and all included RS pairs

Basis Set	n	n_{shell}	μ_q (μs)	σ_q/μ_q	σ_p/μ_p
STO-3G	16	10	32	0.5	0.6
6-31G*	42	20	14	1.4	0.8
cc-pVDZ	58	28	11	3.7	1.2
cc-pVTZ	144	54	10	2.0	1.1
cc-pVQZ	290	88	23	2.6	1.5

The ratio of the standard deviation to the mean tends to increase with improvement of the basis set, although the cc-pVDZ set is an exception. Note that the largest μ_q value is found for the STO-3G basis set; this basis set does not contain uncontracted s functions, which significantly increases the cost of evaluating the s integrals and yields a large mean even though there are no high-angular momentum basis functions.

7.2.2.3 Predicted and Measured Efficiencies

Let us use the performance models in Eqs. 7.7 and 7.8 to analyze the parallel performance of algorithms (a) and (b) in more detail. We will consider the parallel performance on a Linux cluster,[2] and we will employ the values from Table 7.1 for the average time and standard deviation for two-electron integral computation as well as the functional form for k given in Eq. 7.11. In Figure 7.2 we illustrate parallel efficiencies predicted by the models for algorithms (a) and (b). Results were obtained using the correlation-consistent triple-ζ basis set cc-pVTZ for the ethane, butane, and octane molecules. For the ratios σ_q/μ_q and σ_p/μ_p, the values 2.0 and 1.1, respectively, were used. It is clear from Figure 7.2 that the efficiency for algorithm (a) decreases much more slowly with the number of processes than that of algorithm (b), and that, for a given problem size, the efficiency for algorithm (a) is significantly higher than for algorithm (b). In addition to the predicted efficiencies, Figure 7.2 also illustrates actual performance data obtained by running the parallel programs for butane. For both algorithms, there is good agreement between the predicted and measured efficiencies, although for algorithm (b) the model predicts efficiencies that are somewhat too low for small process counts. The results demonstrate the validity of the performance models for both algorithms, the predicted efficiencies showing the correct trends and clearly exposing the shortcomings of algorithm (b).

On the basis of the results presented above we conclude that, for a static distribution of work in the parallel computation of two-electron integrals,

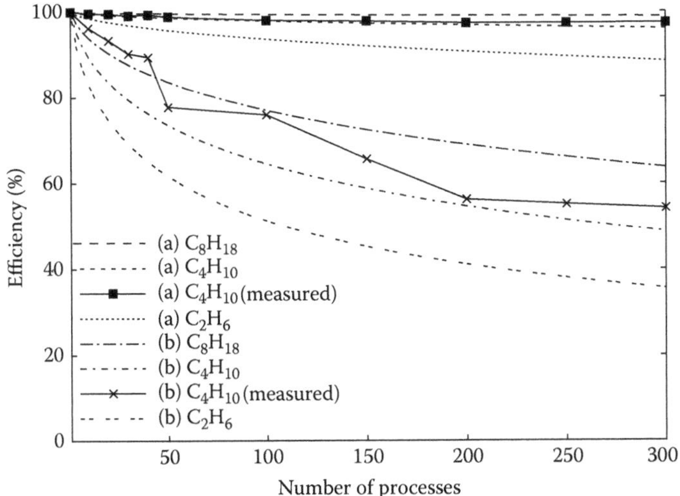

FIGURE 7.2
Predicted and measured parallel efficiencies on a Linux cluster[2] for two-electron integral computation for alkanes employing the cc-pVTZ basis set and using static distribution of shell quartets (a) and shell pairs (b).*

distribution of individual shell quartets can yield very high parallel efficiency, even for large numbers of processes. A static distribution of shell pairs, on the other hand, produces load imbalance that causes the efficiency to decrease, limiting the number of processes that can be utilized. In quantum chemistry applications, however, a static distribution of shell pairs is often employed because it can be advantageous in terms of reducing the memory requirement or the amount of communication (see chapter 9 for an example). Hence, it is of interest to investigate whether the parallel efficiency can be improved when using a pair distribution. This is the topic of the next section in which we will analyze the performance of an algorithm for parallel two-electron integral computation that distributes shell pairs by means of a dynamic manager–worker model.

7.3 Parallel Implementation Using Dynamic Load Balancing

We demonstrated in the previous section that a static distribution of shell pairs in the computation of the two-electron integrals causes a significant performance degradation for large process counts, resulting in a reduced degree of parallelism. For instance, for the butane molecule using the cc-pVTZ

* Since no communication is required by this algorithm, parallelism was simulated by sequentially running batches of integrals. This permitted data to be collected for more processes than the number of available processors in the employed cluster.

basis set, the measured parallel efficiency for the integral computation was around 93% for a process count $p = 20$, about 78% for $p = 50$, and dropped below 60% around $p = 200$. The performance degradation for the static distribution of shell pairs is caused by load imbalance, and we will here discuss an alternative way to distribute shell pairs, employing a dynamic manager–worker distribution to mitigate the load imbalance that arises for the static distribution.

7.3.1 Parallel Algorithm Distributing Shell Pairs

The optimal distribution of tasks (shell pairs) will be the distribution that minimizes the maximum execution time for any process. Finding the optimal distribution can be formulated as a variant of a *bin packing* problem in which a number of items with different weights have to be distributed into a given number of bins so as to minimize the maximum weight of items in a bin. For bin packing problems, it is generally advantageous to sort the items according to decreasing size and to distribute the larger items first. There are a number of commonly used heuristics for providing approximate solutions to bin packing problems, and we will here use the so-called *worst-fit decreasing* heuristic.[4] For the parallel integral computation, this corresponds to sorting the computational tasks according to decreasing size and letting each worker process, once it becomes idle, request the largest remaining task on the manager.

To use this scheme for the parallel integral computation, one process is designated the manager, and a list of tasks (MN shell pairs), sorted according to decreasing size, is created on the manager. The exact sizes (computational times) of the tasks are not known in advance, but we will use the product of the number of basis functions in the M and N shells as a measure for the size of the computational task corresponding to the MN shell pair. These shell pairs will be assigned by the manager to the worker processes by request, one at a time, with the largest tasks distributed first. An algorithm, designated algorithm (c), using this dynamic manager–worker scheme for parallelizing the two-electron integral computation is outlined in Figure 7.3. The manager process only distributes computational tasks and does not compute any integrals itself, and all the computation is done by the worker processes. Each worker process requests individual MN shell pairs from the manager and, after receiving a shell pair, proceeds to compute the $(MN|\overline{RS})$ set of integrals. Upon completing this task, the process requests a new shell pair and so forth until there are no tasks left. At this point, the request for a task will be answered by a message from the manager telling the worker that all tasks have been processed. The manager–worker work distribution scheme has been implemented using point-to-point blocking send and receive operations throughout. Note that this algorithm, like algorithms (a) and (b) in section 7.2, utilizes the full permutational symmetry of the integrals.

```
If (this_proc eq manager)
    Create list of sorted MN shell pairs (descending size)
    int n_MN = n_shell(n_shell + 1)/2
    int requests_remaining = n_MN + p − 1
    int task_index = 0
    While (requests_remaining)
        Receive work request from a worker
        If (task_index < n_MN)
            Send next MN pair from list to worker
            task_index = task_index + 1
        Else
            Send "finished" message to worker
        Endif
        requests_remaining = requests_remaining − 1
    End while
Else
    int finished = 0
    Send request for MN pair to manager
    Receive MN pair or "finished" from manager
    While (!finished)
        For R = 1, M
            If R eq M: S_max = N
            Else: S_max = R
            For S = 1, S_max
                Compute (MN|RS)
            End for
        End for
        Send request for MN pair to manager
        Receive MN pair or "finished" from manager
    End while
Endif
```

FIGURE 7.3
Outline of the parallel algorithm (c) for two-electron integral computation using dynamic manager–worker distribution of shell pairs MN. Indices M, N, R, and S represent shells of basis functions, n_{shell} is the number of shells, p is the process count, and this_proc ($0 \leq$ this_proc $< p$) is the process identifier.

7.3.2 Performance Analysis

Let us consider a performance model for algorithm (c). We first note that, for a manager–worker model in which one process is dedicated to distributing tasks to the others, the maximum efficiency that can be attained with p processes is bounded by $[(p-1)/p] \times 100\%$. Other factors that may contribute to lowering the efficiency are load imbalance on the worker processes and communication overhead.

7.3.2.1 Load Imbalance

The load imbalance resulting from a dynamic distribution of tasks is very difficult to model because the times required for the individual computational tasks are not known in advance. Provided that the number of tasks is much larger than the number of processes, however, it is reasonable to assume that the dynamic task distribution will enable an essentially even distribution of the load. For this to remain true as the number of processes increases, the number of tasks, n_{MN}, must increase proportionally to p. Although this is the same growth rate as obtained for a static work distribution, the actual value for n_{MN} needed for high efficiency for a given process count is much smaller for the dynamic distribution, and the assumption of perfect load balance is therefore adequate for our purposes.

7.3.2.2 Communication Time

Assigning an MN shell pair to a worker requires exchange of two short messages between the manager and a worker: a request from the worker (message length one integer) and a response from the manager (message length two integers). The manager must assign a total of n_{MN} shell pairs to the workers and must also send out a final message to each process when there are no MN pairs left. Using integers of length 4 bytes, the total communication time on the manager then equals $(n_{MN} + p - 1) \times (2 \times \alpha + (4 + 8) \times \beta)$. When a worker receives an assigned MN shell pair, it proceeds to compute all integrals $(MN|\overline{RS})$ for that MN shell pair and all RS shell pairs. This computation of $O(n_{\text{shell}}^2)$ integrals takes considerably longer time than the message exchange with the manager required to get the MN shell pair, and the worker processes therefore spend an insignificant portion of their time on communication. Consider as an example the butane molecule using the correlation-consistent triple-ζ basis set cc-pVTZ. The number of unique MN shell pairs is 4656, and for a calculation with $p = 100$, using $\alpha = 50$ μs and $\beta = 10$ ns, the total communication time on the manager is 0.48 seconds. For this case, the average computational time for a worker process is 2.18 seconds on the employed Linux cluster.[2] Thus, the manager is idle most of the time, and contention on the manager process is unlikely to happen. Additionally, each worker process spends only about $0.48/99 = 4.8$ ms doing communication, which is a negligible fraction (about 0.2%) of the computational time. Therefore, in this case, the communication time can be safely neglected in the performance model.

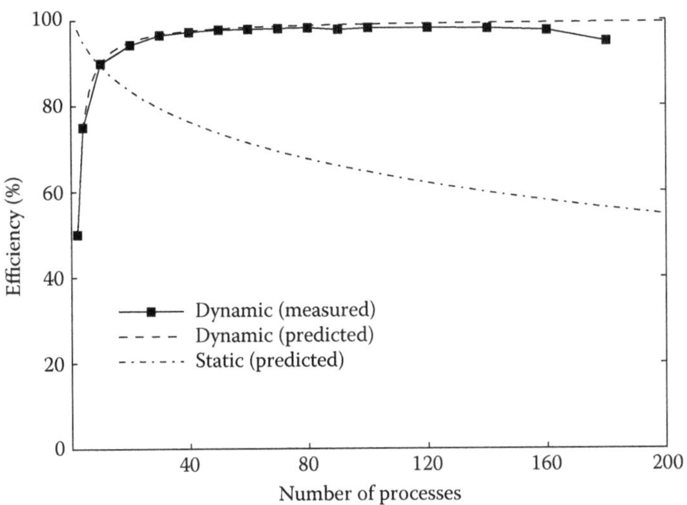

FIGURE 7.4
Predicted and measured parallel efficiencies for two-electron integral computation using dynamic distribution of shell pairs. The predicted efficiency for static distribution of shell pairs is included for comparison. Results were obtained for C_4H_{10} with the cc-pVTZ basis set.

7.3.2.3 Predicted and Measured Efficiencies

With these considerations, we then get the following simple performance model for algorithm (c), expressing the expected efficiency as a function of the number of processes

$$E_c(p) = \frac{p-1}{p}. \tag{7.12}$$

In Figure 7.4, we show performance data for algorithm (c) obtained on a Linux cluster[2] for the butane molecule with the cc-pVTZ basis set. The figure shows both the measured and the predicted efficiency for algorithm (c) and also displays the efficiency predicted for the static pair distribution, algorithm (b), from section 7.2. We first observe that the simple performance model accurately predicts the efficiency for algorithm (c) over a large range of process counts, and that the efficiency increases with the number of processes. The increased efficiency arises because the fraction $(1/p)$ of potential work that is lost by not using the manager for computation of integrals decreases as the number of processes increases. As the process count increases, the efficiency quickly approaches a nearly constant value close to 100%, and algorithm (c) thus exhibits strong scalability, being able to sustain a constant, high efficiency up to nearly 200 processes. As the number of processes grows very large, however, even dynamic load balancing is unable to create an even distribution of work, and the efficiency will eventually decrease.

In summary, we have here analyzed the performance of a dynamic manager–worker model for distributing shell pairs in the parallel computation of two-electron integrals. The analysis clearly demonstrates that the dynamic distribution of shell pairs provides significantly higher parallel efficiency than the static shell pair distribution, except for very small process counts, and that dynamic load balancing enables utilization of a large number of processes with little loss of efficiency.

References

1. Panas, I., J. Almlöf, and M. W. Feyereisen. Ab initio methods for large systems. *Int. J. Quant. Chem.* 40:797–807, 1991.
2. A Linux® cluster consisting of nodes with two single-core 3.06 GHz Intel® Xeon® processors (each with 512 KiB of L2 cache) connected via a 4x Single Data Rate InfiniBand network with a full fat tree topology.
3. Janssen, C. L., I. B. Nielsen, M. L. Leininger, E. F. Valeev, J. Kenny, and E. T. Seidl. The Massively Parallel Quantum Chemistry program (MPQC), version 3.0.0-alpha. Sandia National Laboratories, Livermore, CA, 2007. http://www.mpqc.org.
4. Coffman, Jr., E. G., G. Galambos, S. Martello, and D. Vigo. Bin packing approximation algorithms: Combinatorial analysis. In D.-Z. Du and M. Pardalos (Eds.), *Handbook of Combinatorial Optimization, Supplement Volume A*, pp. 151–207. Boston: Kluwer Academic, 1999.

8

The Hartree–Fock Method

The Hartree–Fock method, also known as the self-consistent field method, is central to quantum chemistry. Incorporating the idea of molecular orbitals, it is a valuable and computationally inexpensive method for providing a qualitative description of the electronic structure of molecular systems. Importantly, the Hartree–Fock method is also the foundation for more sophisticated electronic structure methods that include electron correlation, for instance Møller–Plesset perturbation theory and the coupled-cluster and configuration interaction methods. An efficient Hartree–Fock program is an essential part of a quantum chemistry program suite, and in this chapter we will look at parallel implementation of the Hartree–Fock method. We will first give a brief overview of the computational steps involved in the Hartree–Fock procedure and how these steps can be parallelized. We will then consider in detail the parallel implementation of the formation of the Fock matrix, which is the computationally dominant step. We will discuss two parallel Fock matrix formation algorithms, using replicated and distributed data, respectively.

8.1 The Hartree–Fock Equations

In Hartree–Fock theory, each electron is assigned to a molecular orbital, and the wave function is expressed as a single Slater determinant in terms of the molecular orbitals. For a system with n_{el} electrons the wave function Ψ is then given as

$$\Psi = \frac{1}{\sqrt{n_{\mathrm{el}}!}} \begin{vmatrix} \psi_1(\mathbf{x}_1) & \psi_2(\mathbf{x}_1) & \cdots & \psi_{n_{\mathrm{el}}}(\mathbf{x}_1) \\ \psi_1(\mathbf{x}_2) & \psi_2(\mathbf{x}_2) & \cdots & \psi_{n_{\mathrm{el}}}(\mathbf{x}_2) \\ \vdots & \vdots & \ddots & \vdots \\ \psi_1(\mathbf{x}_{n_{\mathrm{el}}}) & \psi_2(\mathbf{x}_{n_{\mathrm{el}}}) & \cdots & \psi_{n_{\mathrm{el}}}(\mathbf{x}_{n_{\mathrm{el}}}) \end{vmatrix} . \tag{8.1}$$

In Eq. 8.1, ψ_i represents a molecular orbital, and $\mathbf{x}_k$ designates the spatial and spin coordinates of electron k ($\mathbf{r}_k$ and ω_k)

$$\psi_i(\mathbf{x}_k) = \theta_i(\mathbf{r}_k)\sigma(\omega_k) \tag{8.2}$$

where θ_i is a spatial orbital, and σ is a spin function, either α or β. In the following we will consider the closed-shell case, and we will assume that the spin coordinates have been eliminated so that we can formulate the equations in terms of spatial orbitals.

The molecular orbitals are expressed as linear combinations of atomic orbitals ϕ_μ

$$\theta_i(\mathbf{r}) = \sum_{\mu}^{n} C_{\mu i}\phi_\mu(\mathbf{r}) \tag{8.3}$$

where n is the number of atomic orbitals (basis functions), and $C_{\mu i}$ is a molecular orbital coefficient. Expressing the nonrelativistic, time-independent Schrödinger equation using a wave function of this form yields a generalized eigenvalue problem, the Roothaan equations,

$$\mathbf{FC} = \mathbf{SC}\boldsymbol{\epsilon} \tag{8.4}$$

in which $\mathbf{F}$ represents the Fock matrix, $\mathbf{S}$ is the overlap matrix, $\mathbf{C}$ is the matrix of molecular orbital coefficients with elements $C_{\mu i}$, and ϵ is a diagonal matrix of orbital energies. The Fock matrix can be expressed as a sum of a one-electron part $\mathbf{H}^{\text{core}}$ (the core Hamiltonian) and a two-electron part $\mathbf{G}$, and its elements are given as follows in the atomic orbital basis

$$\begin{aligned} F_{\mu\nu} &= H_{\mu\nu}^{\text{core}} + G_{\mu\nu} \\ &= H_{\mu\nu}^{\text{core}} + \sum_{\rho\lambda} D_{\rho\lambda}\left[(\mu\nu|\rho\lambda) - \frac{1}{2}(\mu\lambda|\rho\nu)\right]. \end{aligned} \tag{8.5}$$

The elements $D_{\rho\lambda}$ of the density matrix $\mathbf{D}$ are computed from the molecular orbital coefficients (assumed to be real)

$$D_{\rho\lambda} = 2\sum_{i}^{n_{\text{el}}/2} C_{\rho i}C_{\lambda i} \tag{8.6}$$

where the sum runs over all occupied molecular orbitals i.

The electronic contribution to the Hartree–Fock energy is computed as follows

$$E_{\text{el}} = \frac{1}{2}\sum_{\mu\nu} D_{\mu\nu}(H_{\mu\nu}^{\text{core}} + F_{\mu\nu}) \tag{8.7}$$

and the total Hartree–Fock energy is the sum of the electronic energy and the nuclear repulsion energy: $E_{\text{HF}} = E_{\text{el}} + E_{\text{nuc}}$.

The Roothaan equations are solved for the molecular orbital coefficients **C** and orbital energies ϵ, and the equations must be solved by an iterative procedure because the Fock matrix itself depends on **C**. The most computationally expensive step in the Hartree–Fock procedure is the formation of the two-electron part, **G**, of the Fock matrix; the computation of **G** requires $O(n^4)$ steps, where n is the number of basis functions (with integral screening this cost asymptotically approaches $O(n^2)$). In addition to being the major computational step in the Hartree–Fock procedure, the computation of the **G** matrix is also, by far, the step posing the greatest challenge with respect to efficient parallelization. The **G** matrix is computed from the two-electron integrals, and because of the large computational expense of evaluating these integrals, their full eight-fold permutational symmetry should be utilized, if possible. When using the full symmetry of the integrals $(\mu\nu|\rho\lambda)$, namely $\mu \leftrightarrow \nu$, $\rho \leftrightarrow \lambda$, and $\mu\nu \leftrightarrow \rho\lambda$, the integral $(\mu\nu|\rho\lambda)$ contributes to six Fock matrix elements: $F_{\mu\nu}$ and $F_{\rho\lambda}$ from the first term in the summation on the right hand side of Eq. 8.5 and $F_{\mu\rho}$, $F_{\mu\lambda}$, $F_{\nu\rho}$, $F_{\nu\lambda}$ from the second term in the summation. Consequently, for each integral computed, the corresponding six elements of the density matrix must be available, and six elements of the Fock matrix must be updated with the computed contributions.

8.2 The Hartree–Fock Procedure

Let us briefly consider the steps involved in the Hartree–Fock procedure and how to parallelize these steps. Figure 8.1 outlines the basic computational procedure for solving the Hartree–Fock equations for closed-shell systems. Note that the electronic density matrix **D** is related to the molecular orbital coefficient matrix **C** (for which we are solving the equations) via Eq. 8.6. A guess for **D** is first obtained from, for example, the extended Hückel method or by projecting the density computed with a small basis set into the current basis set. The core Hamiltonian, $\mathbf{H}^{\text{core}}$, is then computed from the kinetic energy and nuclear attraction integrals. This is followed by computation of the overlap matrix for the basis set, **S**, which is diagonalized to obtain the overlap eigenvalues, **s**, and eigenvectors, **U**. Using **s** and **U**, an orthogonal basis is formed, in this case using canonical orthogonalization, which employs the transformation matrix $\mathbf{X} = \mathbf{U}\mathbf{s}^{-1/2}$. At this point, an iterative procedure is begun to determine **C**. In the iterative procedure, the Fock matrix is first formed by computing the density-dependent two-electron part (electron–electron coulomb and exchange contributions), **G**(**D**), and adding it to $\mathbf{H}^{\text{core}}$. The result is transformed into the orthogonal basis, yielding $\mathbf{F}'$, which is then diagonalized to produce the eigenvectors $\mathbf{C}'$ and the eigenvalues ϵ. Next, the eigenvectors $\mathbf{C}'$ are transformed to the original basis, $\mathbf{C} = \mathbf{X}\mathbf{C}'$. Finally, $\mathbf{C}_{\text{occ}}$, which contains the columns of **C** corresponding to the occupied orbitals, is used to compute the new **D**. To accelerate convergence, this density is typically

Guess $\mathbf{D}$

Compute $\mathbf{H}^{\text{core}}$

Compute $\mathbf{S}$

Diagonalize $\mathbf{S}$ to get $\mathbf{s}$ and $\mathbf{U}$

Form $\mathbf{X} = \mathbf{U}\mathbf{s}^{-1/2}$

Iterate until converged:

> Form $\mathbf{F} = \mathbf{H}^{\text{core}} + \mathbf{G}(\mathbf{D})$
>
> Form $\mathbf{F}' = \mathbf{X}^{\dagger}\mathbf{F}\mathbf{X}$
>
> Diagonalize $\mathbf{F}'$ to get ϵ and $\mathbf{C}'$
>
> Form $\mathbf{C} = \mathbf{X}\mathbf{C}'$
>
> Form $\mathbf{D} = 2\mathbf{C}_{\text{occ}}\mathbf{C}_{\text{occ}}^{\dagger}$
>
> Extrapolate $\mathbf{D}$

FIGURE 8.1
An outline of the Hartree–Fock procedure. The Fock matrix $\mathbf{F}$ is a sum of a one-electron part, $\mathbf{H}^{\text{core}}$, and a density-dependent two-electron part, $\mathbf{G}(\mathbf{D})$, where $\mathbf{D}$ is the electronic density matrix. $\mathbf{S}$ is the overlap matrix with eigenvalues $\mathbf{s}$ and eigenvectors $\mathbf{U}$, and $\mathbf{F}'$ is the orthogonalized Fock matrix with eigenvalues ϵ and eigenvectors $\mathbf{C}'$. $\mathbf{C}_{\text{occ}}$ is the portion of the molecular orbital matrix $\mathbf{C}$ corresponding to the occupied orbitals.

extrapolated to obtain the new guess density for the next iteration, and this extrapolation requires operations such as matrix addition and matrix–scalar products.

The major components of the Hartree–Fock procedure, listed roughly in order of decreasing computational requirements, are:

1. Fock matrix formation
2. Matrix diagonalization
3. Matrix multiplication
4. Computation of $\mathbf{H}^{\text{core}}$ and $\mathbf{S}$ matrix elements
5. Matrix addition
6. Matrix-scalar product

In a parallel Hartree–Fock program, these steps can be performed with either replicated or distributed data. When distributed data is used, the matrices are blocked in such a way that all the elements corresponding to the basis functions within a given pair of shells are always found on the same processor. This grouping of the elements by shells is used because integral libraries, for computational efficiency, compute full shell blocks of integrals including all the functions within the involved shells. For the parallel computation of $\mathbf{H}^{\text{core}}$

and **S**, shell pairs are distributed among the processes, and each process computes the parts of these matrices corresponding to the assigned shell pairs. If a replicated data distribution is used, an all-to-all broadcast of these individual contributions will provide the full matrix on each node. The linear algebra operations (matrix diagonalization, multiplication, addition, and scalar product) can be performed with libraries such as ScaLAPACK[1] or Global Arrays.[2] The parallel formation of the Fock matrix, which is the computationally most demanding step, is discussed in detail in sections 8.3 and 8.4 using a replicated data and a distributed data approach, respectively. Replicating all the data allows for a simple parallelization scheme requiring only a single communication step at the end. The use of fully distributed data is significantly more complicated, but it eliminates the possible $O(n^2)$ memory bottleneck of the replicated data approach and creates the potential for treating larger systems.

8.3 Parallel Fock Matrix Formation with Replicated Data

In Figure 8.2, we show the outline of an algorithm for parallel formation of the Fock matrix (two-electron part only) using replicated Fock and density matrices. The algorithm is integral-direct, computing the two-electron integrals on the fly instead of storing them. Each process computes a subset of the two-electron integrals and updates the Fock matrix with the contributions arising from these integrals. Work is assigned to processes by distributing unique atom quartets $ABCD$ ($A \geq B, C \geq D, AB \geq CD$), and letting each process compute the subset of the integrals $(\mu\nu|\rho\lambda)$ for which μ, ν, ρ, and λ are basis functions on atoms A, B, C, and D, respectively; because basis functions are grouped into shells for the purpose of integral computation, this corresponds to computing the integrals $(MN|RS)$ for all shells M, N, R, and S on the atoms A, B, C, and D, respectively.

Processes request tasks (atom quartets) by calling the function get_quartet, which has been implemented in both a dynamic and a static version. The dynamic work distribution uses a manager–worker model with a manager process dedicated to distributing tasks to the other processes, whereas the static version employs a round-robin distribution of tasks. When the number of processes is small, the static scheme achieves the best parallel performance because the dynamic scheme, when run on p processes, uses only $p - 1$ processes for computation. As the number of processes increases, however, the parallel performance for the dynamic task distribution surpasses that of the static scheme, whose efficiency is reduced by load imbalance. With the entire Fock and density matrix available to every process, no communication is required during the computation of the Fock matrix other than the fetching of tasks in the dynamic scheme. After all $ABCD$ tasks have been processed, a global summation is required to add the contributions to the Fock matrix from all processes and send the result to every process.

```
While (get_quartet(A,B,C,D))
    For M ∈ shells on atom A
        For N ∈ shells on atom B
            For R ∈ shells on atom C
                For S ∈ shells on atom D
                    Compute (MN|RS)
                    Update F blocks F_MN, F_RS, F_MR, F_MS, F_NR, F_NS
                            using (MN|RS) and D blocks
                            D_RS, D_MN, D_NS, D_NR, D_MS, D_MR
                End for
            End for
        End for
    End for
End while
Global summation of Fock matrix contributions from all processes
```

FIGURE 8.2
Outline of a parallel algorithm for Fock matrix formation using replicated Fock and density matrices. A, B, C, and D represent atoms; M, N, R, and S denote shells of basis functions. The full integral permutational symmetry is utilized. Each process computes the integrals and the associated Fock matrix elements for a subset of the atom quartets, and processes request work (in the form of atom quartets) by calling the function get_quartet. Communication is required only for the final summation of the contributions to **F**, or, when dynamic task distribution is used, in get_quartet.

The employed grouping together of all the shells on an atom creates larger tasks than if individual shell quartets were distributed. This grouping will tend to worsen load imbalance for the static scheme because the number of tasks is smaller, but the parallel efficiency of the dynamic task distribution will be largely unaffected. The grouping of shells will be advantageous, however, in the distributed data algorithm discussed in section 8.4 because it reduces the communication requirement.

Let us develop a simple performance model for the replicated data algorithm. We will ignore load imbalance in the model and assume that the computation time can be expressed simply as

$$t_{\text{comp}}(p) = \frac{n_{\text{int}}\gamma_{\text{int}}}{p} \tag{8.8}$$

where n_{int} is the number of integrals to be computed, γ_{int} is the average time required to compute an integral, and p is the number of processes. We have here ignored the local update of the Fock matrix elements, which takes a negligible amount of time. The communication overhead is the time required for the global summation of the Fock matrix contributions from all processes; this summation can be performed with an all-reduce operation, and, using Rabenseifner's algorithm (Eq. 3.8), the communication time is

$$t_{\text{comm}}(p) = 2\log_2 p\alpha + \frac{p-1}{p}n^2(\beta + \gamma/2) \tag{8.9}$$

where $\alpha, \beta,$ and γ represent the latency, inverse bandwidth, and floating point operation rate, respectively, and n is the number of basis functions; note that the Fock matrix is symmetric so only about $n^2/2$ elements are stored. Values for α typically lie in the microseconds range, and β and γ will usually be in the nanosecond range; because γ_{int} (Eq. 8.8) typically is on the order of a microsecond and n_{int} is $O(n^4)$ (or $O(n^2)$ with screening), the communication time for this algorithm is negligible relative to the computation time, and the performance model, then, simply predicts linear speedups, $S(p) = p$.

When using this algorithm in a Hartree–Fock program, where the Fock matrix must be computed in each iteration, superlinear speedups are possible, however. With a static task distribution, the Fock matrix formation algorithm can store a number of the computed two-electron integrals and reuse them in each iteration instead of recomputing them: a given process needs the same subset of the integrals in every iteration and can therefore compute all of the required integrals in the first iteration and store a number of them in memory for reuse in later iterations. This integral storage has consequences for the parallel performance of the algorithm. If we assume that each process can store m integrals, the process will compute n_{int}/p integrals in the first iteration and $n_{\text{int}}/p - m$ integrals (when $n_{\text{int}}/p \geq m$) in each of the following iterations. The speedup can then be expressed as

$$\begin{aligned} S(p) &= \frac{n_{\text{int}} + (n_{\text{iter}} - 1)(n_{\text{int}} - m)}{n_{\text{int}}/p + (n_{\text{iter}} - 1)(n_{\text{int}}/p - m)} \\ &= \frac{n_{\text{int}}n_{\text{iter}} - m(n_{\text{iter}} - 1)}{n_{\text{int}}n_{\text{iter}}/p - m(n_{\text{iter}} - 1)} \\ &= p \times \frac{n_{\text{int}}n_{\text{iter}} - m(n_{\text{iter}} - 1)}{n_{\text{int}}n_{\text{iter}} - pm(n_{\text{iter}} - 1)} \end{aligned} \tag{8.10}$$

where n_{iter} represents the number of iterations. The second term in the denominator in Eq. 8.10 gives rise to an upward concave speedup curve, yielding superlinear speedups. Superlinear speedups are possible only when the algorithm is used in an iterative procedure and each process allocates some memory for integral storage. The results presented below pertain to a single computation of the Fock matrix, in which case integral reuse is not possible, and the ideal speedups will be the usual linear speedups, $S(p) = p$. In section 5.4, superlinear speedups are shown for a Hartree–Fock program that uses the above Fock matrix formation algorithm and employs integral storage.

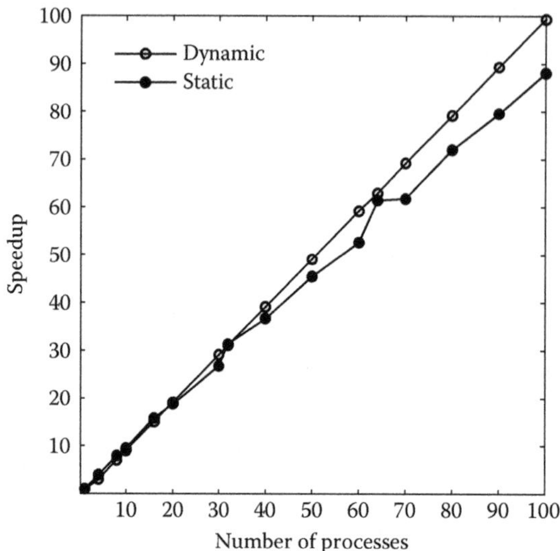

FIGURE 8.3
Speedups for parallel Fock matrix formation using replicated density and Fock matrices. Speedups were obtained on a Linux cluster[3] for the uracil dimer with the aug-cc-pVTZ basis set and were computed relative to single-process timings using measured wall times.

In Figure 8.3 we show speedups for the replicated data algorithm, measured for the uracil dimer, $(C_4N_2O_2H_4)_2$, using the augmented correlation-consistent triple-ζ basis set aug-cc-pVTZ (920 basis functions). The algorithm displays nearly ideal speedups when using a dynamic task distribution, achieving a speedup of 99 when running on 100 processes. The speedups for the static distribution are high as well, although a bit lower than for the dynamic case because load imbalance lowers the efficiency for larger process counts; the static distribution yields a speedup of 88 on 100 processes.

Parallel Fock matrix computation using replicated density and Fock matrices is easy to implement and achieves high parallel efficiency. Using replicated matrices, however, may create a memory bottleneck: the Fock and density matrices are both, nominally, of size $O(n^2)$, and keeping a copy of the entire density and Fock matrix for every process may not be possible when the number of basis functions is very large. This memory bottleneck can be avoided by distributing the matrices, and we will explore this approach in the next section.

8.4 Parallel Fock Matrix Formation with Distributed Data

To avoid a potential memory bottleneck in parallel Hartree–Fock computations for very large systems, the density and Fock matrices must be distributed, and in this section we will investigate a distributed data parallel

```
Boolean got_quartet = get_quartet(A',B',C',D')
If (got_quartet)
    begin_prefetch_blocks(A',B',C',D')
Endif
While (got_quartet)
    A = A'; B = B'; C = C'; D = D'
    got_quartet = get_quartet(A',B',C',D')
    finish_prefetch_blocks(A,B,C,D)
    If (got_quartet)
        begin_prefetch_blocks(A',B',C',D')
    Endif
    For M ∈ shells on atom A
        For N ∈ shells on atom B
            For R ∈ shells on atom C
                For S ∈ shells on atom D
                    Compute (MN|RS)
                    Update F blocks: F_MN, F_RS, F_MR, F_MS, F_NR, F_NS
                                using (MN|RS) and D blocks
                                D_RS, D_MN, D_NS, D_NR, D_MS, D_MR
                End for
            End for
        End for
    End for
    accumulate_blocks(A,B,C,D)
    flush_block_cache(A,B,C,D)
End while
```

FIGURE 8.4
Outline of a parallel algorithm for Fock matrix formation using distributed Fock and density matrices. A, B, C, and D represent atoms, M, N, R, and S denote shells of basis functions, and only unique integrals are computed.

Fock matrix formation algorithm. Again, we will discuss only the computation of the two-electron part of the Fock matrix.

An outline of the algorithm is shown in Figure 8.4. Like the replicated data algorithm from section 8.3, the distributed data algorithm is integral-direct, and the computational tasks to be distributed are atom quartets $ABCD$. The tasks can be distributed statically using a round-robin scheme or dynamically by means of a manager–worker model. Data distribution is accomplished by distributing atom blocks F_{AB} and D_{AB} of the Fock and density matrices across all processes; the blocks F_{AB} and D_{AB} include all matrix elements $F_{\mu\nu}$ and $D_{\mu\nu}$ with μ and ν representing basis functions on atoms A and B, respectively. Using distributed Fock and density matrices necessitates communication throughout the computation of the Fock matrix and makes efficient parallelization significantly more challenging. Because the full permutational symmetry of the integrals is utilized, each atom quartet of integrals $(AB|CD)$ contributes to six atom blocks of the Fock matrix (F_{AB}, F_{CD}, F_{AC}, F_{AD}, F_{BC}, F_{BD}); to compute these contributions, six blocks of the density matrix (D_{CD}, D_{AB}, D_{BD}, D_{BC}, D_{AD}, D_{AC}) are required, and these blocks must be fetched from the processes that store them. To improve performance, prefetching of density matrix blocks is employed, so that a process will not need to wait idly for data to arrive. When the contributions to the Fock matrix from an atom quartet of integrals have been computed, they are sent to the processes that store the corresponding blocks of the Fock matrix. The prefetching of density matrix blocks and the sending of computed Fock matrix contributions to the processes where they belong are effectuated by a one-sided message passing scheme; this scheme is implemented via multi-threading using a version of MPI that provides full support for multiple threads. By using one-sided communication, it is possible to overlap computation and communication, and, hence, a process may continue to perform computation while, at the same time, a communication thread within the same process is active exchanging data with other processes. Use of a one-sided communication scheme is necessary to achieve high parallel performance in the distributed data Fock matrix computation because the varying sizes of the atom quartets, combined with the need to retrieve or send out density and Fock matrix blocks, result in an irregular communication pattern involving frequent message exchanges.

The algorithm, as outlined in Figure 8.4, proceeds as follows. Initially, a process calls the function `get_quartet` to fetch a computational task, namely an atom quartet $A'B'C'D'$. If successful, the process then calls the function `begin_prefetch_blocks` to start prefetching the required blocks of the density matrix ($D_{C'D'}$, $D_{A'B'}$, $D_{B'D'}$, $D_{B'C'}$, $D_{A'D'}$, $D_{A'C'}$). A `while` loop is then entered, and inside this loop, a new atom quartet is first fetched; the prefetching of the density matrix blocks for the previous atom quartet is then completed (by calling `finish_prefetch_blocks`), and prefetching of density matrix blocks for the new atom quartet is initiated. While the prefetching for the new quartet is taking place, the integrals in the old quartet are computed and contracted with the prefetched blocks of the density matrix (Eq. 8.5) inside the loop over the shells M, N, R, and S on the atoms A, B, C, and D to produce the corresponding Fock matrix contributions. When these contributions have been computed for the entire current atom quartet, they are sent to the processes where they are to be stored and added into the local

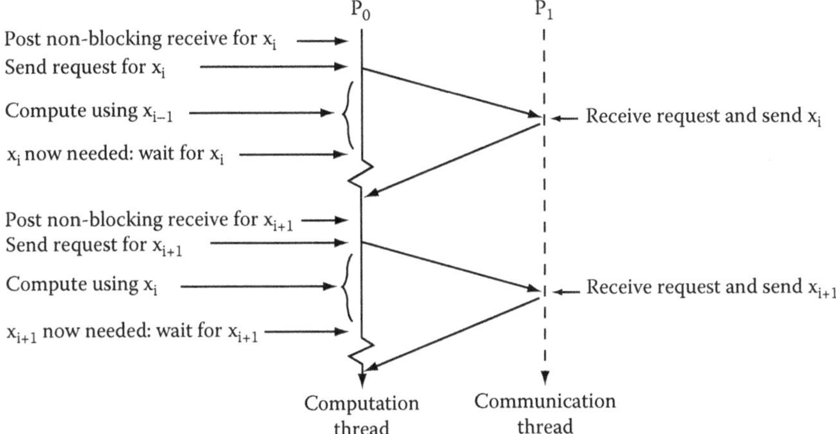

FIGURE 8.5
Prefetching of data by process P_0 from process P_1 in the parallel Fock matrix computation with distributed data. P_0 and P_1 each run a computation thread and a communication thread; only the threads involved in the data exchange, namely the computation thread within P_0 and the communication thread within P_1, are shown. The zigzag pattern on the computation thread represents (potential) idle time.

Fock matrix blocks held by those processes. This is accomplished by calling the function `accumulate_blocks`, and after thus accumulating the locally computed contributions to the Fock matrix remotely, the blocks of the density matrix used to compute these contributions are cleared (by calling the function `flush_block_cache`) before the next batch is fetched.

As discussed in section 7.3, it is advantageous from a load balancing perspective, when using a dynamic task distribution, to process computationally expensive blocks first; however, if we are otherwise careless about the order in which tasks are processed, it is possible for all of the tasks to simultaneously request the same density block, resulting in a communication bottleneck for the process holding that block. Thus, even when dynamic task distribution is used, some degree of randomization is employed in the task ordering.

The multi-threaded approach used in the algorithm is illustrated in Figure 8.5. Each process spawns a separate communication thread that handles all requests from remote processes without blocking the work done by the process's main computation thread. Two types of requests can be sent to the communication threads: requests for a specific block of the density matrix to be sent out and requests to add data into a locally stored Fock matrix block. Figure 8.5 only shows the fetching of the density matrix elements, and the case illustrated involves only two processes, P_0 and P_1. The illustrated data exchange involves P_0 requesting data from P_1, and only the threads involved, namely the computation thread within P_0 and the communication thread within P_1, are shown in the figure. Note that the computation thread in P_1 is not blocked by the data processing handled by P_1's communication thread: communication and computation can progress simultaneously. As shown in the figure, the computation thread within P_0 first initiates a prefetch of a block of the density matrix by sending a request to P_1, whose communication thread receives this request and starts sending the requested block back to P_0. Meanwhile, after issuing the prefetch, the computation thread in P_0 continues to do computation until, at some later point, it needs to process the data requested from P_1. At this point, the computation thread issues a `wait` call, which causes the computation to pause until the requested data have arrived. Ideally, the data should already have arrived at this time, so only a short interruption of execution in P_0's computation thread is required to ascertain that the requested block is available.

Let us look at the communication requirement for this algorithm. For each atom quartet, six blocks of the density matrix must be fetched, and the corresponding six blocks of the Fock matrix must be sent where they belong. The average size of these blocks is approximately $(n/n_{\text{atom}})^2$. The communication requirement per process then becomes

$$t_{\text{comm}}(p) = \frac{n_{\text{quartet}}}{p}[12\alpha + 12(n/n_{\text{atom}})^2\beta] \tag{8.11}$$

where n_{quartet}/p represents the number of atom quartets handled by each process, and n_{quartet} is given as

$$n_{\text{quartet}} \approx \begin{cases} n_{\text{atom}}^4/8 & \text{without screening} \\ k^2 n_{\text{atom}}^2/8 & \text{with screening.} \end{cases} \tag{8.12}$$

We have here assumed that, with integral screening, each of the basis functions on an atom, on average, interact with the basis functions on k other atoms. Note that fetching entire atom blocks of the density at a time and sending out computed Fock matrix contributions by atom blocks reduces the total amount of data to be communicated by a factor of $(n/n_{\text{atom}})^2$ (without screening) relative to the communication requirement resulting from sending and fetching individual matrix elements for each two-electron integral.

Having determined the communication requirement, we can work out a performance model for the algorithm. The amount of computation per process is the same as for the replicated data algorithm, namely

$$t_{\text{comp}}(p) = \frac{n_{\text{int}}\gamma_{\text{int}}}{p} \tag{8.13}$$

where n_{int} denotes the number of integrals to be computed, and γ_{int} represents the average time required to compute an integral. The value of n_{int} is given as

$$n_{\text{int}} \approx \begin{cases} n^4/8 & \text{without screening} \\ k^2 n^4/(8n_{\text{atom}}^2) & \text{with screening.} \end{cases} \tag{8.14}$$

With screening performed at the atom level, n_{int} is computed as the product of the number of atom quartets, $\frac{1}{8}(kn_{\text{atom}})^2$, and the average atom quartet size, $(n/n_{\text{atom}})^4$.

Using the earlier expressions for the computation time and the communication time, we can derive the efficiency for the algorithm. First, note that the efficiency can be expressed as follows

$$
\begin{aligned}
E(p) &= \frac{S(p)}{p} \\
&= \frac{t_{\text{comp}}(1)}{p\left[t_{\text{comp}}(1)/p + t_{\text{comm}}(p)\right]} \\
&= \frac{1}{1 + t_{\text{comm}}(p)/t_{\text{comp}}(p)}
\end{aligned}
\tag{8.15}
$$

where we have used the relation $t_{\text{comp}}(p) = t_{\text{comp}}(1)/p$, which assumes that the computation steps are perfectly parallelized. Thus, the efficiency is a function of the ratio of the communication time to the computation time. This ratio can be derived from Eqs. 8.11–8.14.

$$
\frac{t_{\text{comm}}(p)}{t_{\text{comp}}(p)} = \left(\frac{n_{\text{atom}}}{n}\right)^4 \frac{12\alpha + 12(n/n_{\text{atom}})^2\beta}{\gamma_{\text{int}}}. \tag{8.16}
$$

Note that this ratio (and, hence, the parallel efficiency) is not affected by the use of integral screening. Because the value for the ratio $t_{\text{comm}}(p)/t_{\text{comp}}(p)$ is independent of p, the efficiency is independent of p, and the algorithm is strongly scalable. Scalability is a particularly important property for a distributed data algorithm whose advantage relative to the replicated data version is a reduced memory requirement per process: to fully utilize the distributed data algorithm's potential for handling larger systems, the number of processes should be large, and the algorithm, therefore, must be able to run efficiently also for large process counts.

If values for α, β, γ_{int}, and n/n_{atom} are known, or can be estimated, the efficiency provided by the algorithm can be computed. Typical values for α and β for current high-performance networks (see Table 5.1) are α = 30 μs and β = 14 ns/word (using 8 byte words), and the time required to compute an integral, γ_{int}, is on the order of a microsecond on a current state-of-the-art microprocessor for basis sets of double-ζ plus polarization quality; we will use a value of 0.5 μs for γ_{int}, obtained on the Linux cluster[3] used for testing the algorithm. For a double-ζ plus polarization basis set, assuming a roughly equal number of hydrogen and first-row atoms, the number of basis functions per atom is $n/n_{\text{atom}} \approx 10$. Using these values, and substituting Eq. 8.16 into Eq. 8.15, we get a parallel efficiency $E(p) \approx 93\%$ for the distributed data Fock matrix formation algorithm. Note that in addition to depending on the number of basis functions per atom, the predicted efficiency varies considerably with the relative values of γ_{int} and the parameters α and β. For instance, on a network with lower performance, assuming a five-fold increase in α and β,

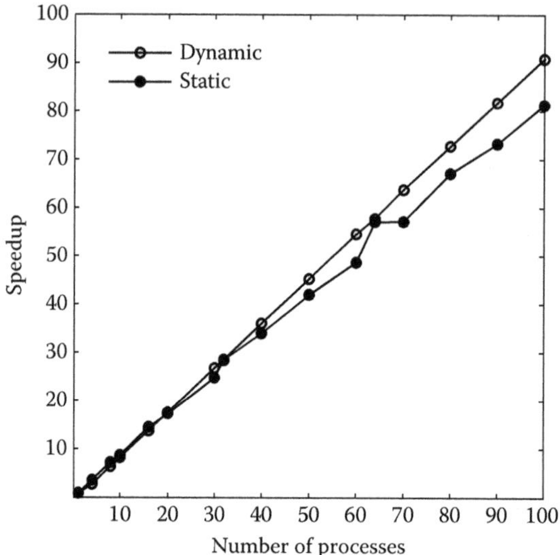

FIGURE 8.6
Speedups for parallel Fock matrix formation using fully distributed density and Fock matrices. Speedups were obtained using one compute thread per node on a Linux cluster[3] for the uracil dimer with the aug-cc-pVTZ basis set. Speedups were computed relative to single-process timings using measured wall times.

the predicted efficiency would be reduced to about 73%; a decrease in processor performance, on the other hand, such as increasing γ_{int} by a factor of five, would increase the parallel efficiency to around 99%.

Speedup curves for the distributed data algorithm running on a Linux cluster[3] are shown in Figure 8.6. Speedups were measured for the same case used for the replicated data algorithm in Figure 8.3, and an ideal speedup curve corresponds to $S(p) = p$. Timings were measured with both a dynamic and a static task allocation, and the dynamic scheme, which provides a more even work distribution, yields better parallel performance for large process counts, achieving a speedup of 91 for 100 processes compared with 81 for the static distribution. The speedups for both the dynamic and static task distributions in Figure 8.6 are somewhat lower than their counterparts in the replicated data algorithm, Figure 8.3, because the communication overhead in the distributed data algorithm is nonnegligible. As predicted, the distributed data algorithm is scalable, yielding linear speedup curves and maintaining a nearly constant efficiency as the number of processes increases.

To further investigate the scalability of the Fock matrix formation algorithms as well as the effect of running multiple compute threads on a node, a series of runs were performed using two compute threads (and one communication thread) per node, enabling computations to be performed with up to 200 compute threads; apart from the number of compute threads per node, the test case was identical to the one used above. The resulting speedups are

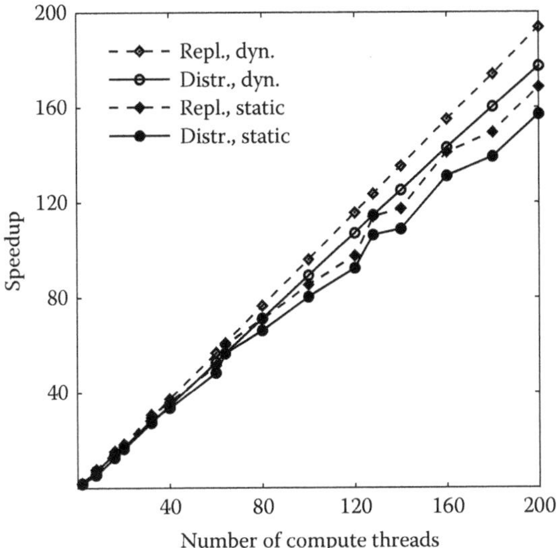

FIGURE 8.7
Speedups for parallel Fock matrix formation for distributed and replicated data algorithms when running two compute threads per node. Speedups were obtained on a Linux cluster[3] for the uracil dimer with the aug-cc-pVTZ basis set and were computed relative to single-node timings with one compute thread, using measured wall times.

shown in Figure 8.7 for both the replicated and distributed data algorithms. For a given total number of compute threads, the speedups measured using two compute threads per node are slightly lower, about 1–3%, than those obtained for a single compute thread per node (cf. Figures 8.3 and 8.6). This small decline in performance is caused by competition for shared resources, including communication and memory bandwidth, between the two compute threads on each node; in the distributed data case, the compute threads also compete with the communication thread for the two processors available on each node. When using a static task distribution, load imbalance causes somewhat jagged speedup curves and a slightly decreasing efficiency, but with dynamic load balancing, both algorithms are capable of providing sustained high performance as the number of compute threads increases.

8.5 Further Reading

For a detailed discussion of Hartree–Fock theory, see, for instance, Szabo and Ostlund.[4] Many parallel self-consistent field implementations have been presented in the literature; for a review of some of the early work in this field, see Harrison and Shepard.[5] Several massively parallel, distributed data self-consistent field algorithms have been implemented. For example, Colvin et al.[6]

developed a direct algorithm using a static work distribution together with a systolic loop communication pattern, and Harrison et al.[7] employed dynamic load balancing and distributed data by means of global arrays accessed via a one-sided communication scheme. A semidirect approach used in conjunction with a dynamic load balancing strategy was investigated by Lindh et al.,[8] and Mitin et al.[9] have implemented disk-based and semidirect parallel Hartree–Fock programs with compression of the integrals stored on disk.

References

1. Information on the ScaLAPACK library can be found at http://www.netlib.org/scalapack/scalapack_home.html.
2. Nieplocha, J., R. J. Harrison, and R. J. Littlefield. Global arrays: A portable "shared-memory" programming model for distributed memory computers. In *Proceedings of the 1994 Conference on Supercomputing*, pp. 340–349. Los Alamitos: IEEE Computer Society Press, 1994.
3. A Linux® cluster consisting of nodes with two single-core 3.6 GHz Intel® Xeon® processors (each with 2 MiB of L2 cache) connected with a 4x Single Data Rate InfiniBand™ network using Mellanox Technologies MT25208 InfiniHost™ III Ex host adaptors and a Topspin 540 switch. The InfiniBand host adaptors were resident in a PCI Express 4x slot, reducing actual performance somewhat compared to using them in an 8x slot. MPICH2 1.0.5p4 was used with TCP/IP over InfiniBand IPoIB because this configuration provided the thread-safety required for some applications.
4. Szabo, A., and N. S. Ostlund. *Modern Quantum Chemistry*, 1st revised edition, chapter 3. New York: McGraw-Hill, 1989.
5. Harrison, R. J., and R. Shepard. Ab initio molecular electronic structure on parallel computers. *Ann. Rev. Phys. Chem.* 45:623–658, 1994.
6. Colvin, M. E., C. L. Janssen, R. A. Whiteside, and C. H. Tong. Parallel direct SCF for large-scale calculations. *Theor. Chim. Acta* 84:301–314, 1993.
7. Harrison, R. J., M. F. Guest, R. A. Kendall, D. E. Bernholdt, A. T. Wong, M. Stave, J. L. Anchell, A. C. Hess, R. J. Littlefield, G. L. Fann, J. Nieplocha, G. S. Thomas, D. Elwood, J. T. Tilson, R. L. Shepard, A. F. Wagner, I. T. Foster, E. Lusk, and R. Stevens. Toward high-performance computational chemistry: II. A scalable self-consistent field program. *J. Comp. Chem.* 17:124–132, 1996.
8. Lindh, R., J. W. Krogh, M. Schütz, and K. Hirao. Semidirect parallel self-consistent field: The load balancing problem in the input/output intensive self-consistent field iterations. *Theor. Chem. Acc.* 110:156–164, 2003.
9. Mitin, A. V., J. Baker, K. Wolinski, and P. Pulay. Parallel stored-integral and semidirect Hartree-Fock and DFT methods with data compression. *J. Comp. Chem.* 24:154–160, 2003.

9

Second-Order Møller–Plesset Perturbation Theory

Second-order Møller–Plesset (MP2) perturbation theory is among the simplest and most widely used quantum chemical methods for incorporating electron correlation. Although MP2 theory is computationally less expensive than most other correlated electronic structure methods, its computational cost scales steeply with the molecular size. For conventional implementations of MP2 theory, which are based on canonical orbitals obtained in a Hartree–Fock calculation, the scaling is formally $O(N^5)$, where N is the size of the molecule. Parallel implementation of MP2 theory, therefore, is an important step towards extending the applicability of the method to larger molecular systems. Another potential bottleneck in applications of conventional MP2 theory is the rather large storage requirement, which arises in the transformation of the two-electron integrals from the atomic orbital to the molecular orbital basis. To mitigate this bottleneck, a parallel implementation of MP2 theory should use a distributed data model in which the integral arrays are distributed across processes.

In the following we will discuss scalar and parallel implementations of canonical closed-shell MP2 theory. We will present a scalar MP2 energy algorithm and show two different parallel implementations that both employ distributed data but differ in their level of sophistication. The first parallel implementation is readily obtained from the scalar code, requiring only small modifications, whereas the second parallel algorithm employs a more sophisticated asynchronous message-passing scheme and is designed to be scalable.

9.1 The Canonical MP2 Equations

The MP2 energy is the sum of the Hartree–Fock energy and the MP2 correlation energy

$$E_{\mathrm{MP2}} = E_{\mathrm{HF}} + E_{\mathrm{MP2}}^{\mathrm{corr}} \tag{9.1}$$

and the MP2 correlation energy for a closed-shell system can be expressed as[1,2]

$$E_{\mathrm{MP2}}^{\mathrm{corr}} = \sum_{ijab} \frac{(ia|jb)[2(ia|jb) - (ib|ja)]}{\epsilon_i + \epsilon_j - \epsilon_a - \epsilon_b} \tag{9.2}$$

where i, j and a, b represent occupied and unoccupied (virtual) spatial molecular orbitals, respectively, ϵ_m denotes a molecular orbital energy, and $(ia|jb)$ is a two-electron integral defined as

$$(ia|jb) = \sum_{\mu\nu\lambda\sigma} C_{\mu i} C_{\nu a} C_{\lambda j} C_{\sigma b} (\mu\nu|\lambda\sigma) \tag{9.3}$$

$$(\mu\nu|\lambda\sigma) = \int \phi_\mu(\mathbf{r}_1)\phi_\nu(\mathbf{r}_1) r_{12}^{-1} \phi_\lambda(\mathbf{r}_2)\phi_\sigma(\mathbf{r}_2) d\mathbf{r}_1 d\mathbf{r}_2. \tag{9.4}$$

The coefficients $C_{\mu m}$ are the molecular orbital coefficients obtained in the Hartree–Fock procedure, and ϕ_μ is an atomic orbital. The two-electron integral transformation (Eq. 9.3) is usually performed as four separate quarter transformations

$$(i\nu|\lambda\sigma) = \sum_{\mu} C_{\mu i} (\mu\nu|\lambda\sigma) \tag{9.5}$$

$$(i\nu|j\sigma) = \sum_{\lambda} C_{\lambda j} (i\nu|\lambda\sigma) \tag{9.6}$$

$$(ia|j\sigma) = \sum_{\nu} C_{\nu a} (i\nu|j\sigma) \tag{9.7}$$

$$(ia|jb) = \sum_{\sigma} C_{\sigma b} (ia|j\sigma) \tag{9.8}$$

which reduces the overall computational complexity of the transformation from $O(N^8)$ (from Eq. 9.3) to $O(N^5)$, where N represents the size of the system. For the integral transformation as written above, the computational complexity of each of the four quarter transformations is $O(on^4)$, $O(o^2n^3)$, $O(o^2vn^2)$, and $O(o^2v^2n)$, respectively, where o is the number of occupied orbitals, v is the number of virtual orbitals, and n denotes the number of basis functions. Transforming to occupied, rather than virtual, molecular orbitals in the first two quarter transformations results in computational savings and a reduced storage requirement because the number of occupied orbitals is usually significantly smaller than the number of virtual orbitals.

The integral transformation may be performed by first computing all the two-electron integrals in the atomic orbital basis, $(\mu\nu|\lambda\sigma)$, and then doing the four quarter transformations consecutively, completing one quarter transformation before beginning the next. In this case, however, it is necessary to store all the computed two-electron integrals in the atomic orbital basis as well as all the partially transformed integrals generated in each transformation step. This leads to a storage requirement of $O(N^4)$, which becomes a bottleneck as the size of the system grows. To reduce the storage requirement, a direct approach may be employed in which a subset of the two-electron integrals in the atomic orbital basis is computed and partially transformed before computing the next subset of integrals. If MP2 calculations are to be performed for large molecular systems, a direct method will usually be used. The two-electron integral transformation is the most time-consuming step in the computation of the MP2 energy, and the computational efficiency of this step is therefore important for the overall efficiency of the MP2 procedure. To reduce the computational time required for the transformation, the permutational symmetry of the integrals should be exploited to the extent possible. At the same time, the storage requirement should be kept low to prevent storage bottlenecks, and the optimum implementation may entail a compromise between these two requirements.

9.2 A Scalar Direct MP2 Algorithm

An outline of a scalar direct MP2 algorithm is shown in Figure 9.1, and the major features of the algorithm are summarized in Table 9.1. The algorithm is a modified version of a previously published direct MP2 algorithm[3] designed

Loop over I batches

  Loop over R, S shells $(S \le R)$

    Loop over M, N shells $(N \le M)$

      Compute $(MN|RS)$

      Loop over $i \in I$

        $(iN|RS) = (iN|RS) + C_{Mi}(MN|RS)$

        $(iM|RS) = (iM|RS) + C_{Ni}(MN|RS)$

      End i loop

    End M, N loop

    Loop over $i \in I$, $j \le i$, all N

      $(iN|jS) = (iN|jS) + C_{Rj}(iN|RS)$

      $(iN|jR) = (iN|jR) + C_{Sj}(iN|RS)$

    End i, j, N loop

  End R, S loop

  $(ia|jS) = \sum_N C_{Na}(iN|jS)$ $(i \in I$; all $j, a, S, N)$

  $(ia|jb) = \sum_S C_{Sb}(ia|jS)$ $(i \in I$; all $j, a, b, S)$

  $E_{\text{corr}}^{\text{MP2}} = E_{\text{corr}}^{\text{MP2}} + \sum_{ijab}(ia|jb)[2(ia|jb) - (ib|ja)]/(\epsilon_i + \epsilon_j - \epsilon_a - \epsilon_b)$
  $(i \in I$; all $j, a, b)$

End I loop

FIGURE 9.1
A scalar direct MP2 algorithm. R, S, M, and N denote shells of atomic orbitals. To reduce the storage requirement, occupied orbitals, i, can be processed in batches, I.

TABLE 9.1

Comparison of major features of the scalar and parallel MP2 algorithms. The scalar algorithm and the parallel algorithms P1 and P2 are outlined in Figures 9.1, 9.2, and 9.3, respectively. The number of basis functions, occupied, and virtual orbitals are denoted n, o, and v, respectively; n_{shell}, $n_{\text{sh,max}}$, n_i, and p represent, in order, the number of shells, the maximum shell size, the number of occupied orbitals in a batch, and the number of processes. Process utilization is defined as the number of processes that can be efficiently utilized

	Scalar	P1	P2
Symmetry utilization (AO integrals)	4-fold	4-fold	4-fold
Memory requirement per process	$O(on^2)$	$O(on^2/p)$	$O(on^2/p)$
1st quarter transformation	$n_i n n_{sh,max}^2$	$n_i n n_{sh,max}^2$	$n_i n n_{sh,max}^2$
2nd quarter transformation	$n_i on^2$	$n_i on^2/p$	$n_i on^2/p$
3rd quarter transformation	$n_i ovn$	$n_i ovn/p$	$n_i ovn/p$
4th quarter transformation	$n_i ov^2$	$n_i ov^2/p$	$n_i ov^2/p$
Communication per process	N/A	$O(on^3)$	$O(n_{\text{shell}} o^2 n^2/p)$
Type of communication	N/A	Collective	One-sided
Process utilization	1	$O(o^2)$	$O(o^2)$

to have a small memory requirement and yet be computationally efficient. A fully in-core approach is used, employing no disk storage. The algorithm is integral-direct, requiring storage of only a subset of the two-electron integrals at any given time, and two of the three permutational index symmetries, $M \leftrightarrow N$ and $R \leftrightarrow S$, can be exploited in the computation of the two-electron integrals $(MN|RS)$, where M, N, R, and S represent shells of atomic orbitals (basis functions). The memory requirement is only third-order, $O(ovn)$, and is significantly smaller than the fourth-order requirement in a conventional approach. In the integral computation and the first two quarter transformations, the memory requirement is kept small by performing these steps for one RS pair at a time (this is what prohibits utilization of the third index permutation symmetry, $MN \leftrightarrow RS$, in the integral computation). The overall memory requirement is reduced from fourth- to third-order by dividing the occupied orbitals, i, into batches and performing the integral transformation for one batch at time, computing the contribution to the correlation energy from the generated integrals in the current batch before processing the next one. The batch size, n_i, which may range from 1 to the number of occupied orbitals, o, is determined by the available memory, and with n_i occupied orbitals in a batch, the memory requirement becomes $n_i ovn$. Reducing the batch size thus reduces the memory requirement but increases the total amount of computation because the two-electron integrals must be computed in each batch.

The algorithm, as outlined in Figure 9.1, entails an outer loop over batches, I, of occupied orbitals. For each batch, the two-electron integrals $(MN|RS)$ are computed for one shell pair R, S ($S \leq R$) and all M, N pairs ($N \leq M$), and these integrals are then immediately transformed in the first quarter transformation, generating $(iN|RS)$. Because of the restriction $N \leq M$, the computed $(MN|RS)$ integrals must be summed into both $(iN|RS)$ and $(iM|RS)$. The second quarter transformation is performed within the same loop over R and S, and both the atomic orbital integrals, $(MN|RS)$, and the quarter-transformed integrals, $(iN|RS)$, are stored for only one R, S pair at a time. The half-transformed integrals, $(iN|jS)$, and the three-quarter and fully transformed integrals, $(ia|jS)$ and $(ia|jb)$, respectively, are generated for the full range of all pertinent indices except for the occupied index i, which belongs to the current batch of occupied orbitals.

9.3 Parallelization with Minimal Modifications

Parallel implementations of quantum chemical methods are frequently derived from existing serial code, rather than developing the parallel algorithms from scratch. If a serial code already exists, this approach will often be the fastest way to achieve a parallel implementation. The relative ease of implementation, however, generally comes at the expense of a lower parallel efficiency than what could be achieved by designing a parallel algorithm from the beginning. In this section we consider a parallel direct MP2 energy algorithm based on the scalar algorithm presented in section 9.2 and requiring only small changes to the serial code.

The minimum requirements for a reasonably efficient parallel MP2 implementation that can utilize a large number of processes are: distribution of all large data arrays (to avoid storage bottlenecks) and distribution of the two-electron integral computation and transformation over two or more indices (to balance the load). Load balancing in the computation of the two-electron integrals requires distribution of work over at least two indices because the computational tasks are not even-sized. Thus, if the computation of $(MN|RS)$ is distributed over only one shell index, there will be n_{shell} large, but irregularly sized, tasks to distribute, and load balance will be difficult to achieve for large process counts, especially with a static task distribution. Distribution over two shell indices, on the other hand, creates $O(n_{\text{shell}}^2)$ smaller tasks that can more easily be evenly distributed, making it possible to take advantage of more processes. A detailed discussion of load balancing in the computation of the two-electron integrals is given in chapter 7.

The parallel algorithm, P1, developed by doing only minimal modifications to the scalar algorithm from the previous section is shown in Figure 9.2. Parallelization has been achieved using a simple global communication scheme that is straightforward to implement, and the work has been distributed as follows: in the integral transformation, a round-robin distribution

Loop over I batches
    Loop over R, S shells ($S \leq R$)
        Loop over M, N shells ($N \leq M$)
            If my $\underline{MN}$ pair
                Compute $(\underline{MN}|RS)$
                Loop over $i \in I$
                    $(iN|RS) = (iN|RS) + C_{Mi}(\underline{MN}|RS)$
                    $(iM|RS) = (iM|RS) + C_{Ni}(\underline{MN}|RS)$
                End i loop
            Endif
        End M, N loop
        Global summation of current $(iN|RS)$ batch
        Loop over $i \in I$, all j, N
            If my $\underline{ij}$ pair
                $(\underline{i}N|\underline{j}S) = (\underline{i}N|\underline{j}S) + C_{Rj}(iN|RS)$
                $(\underline{i}N|\underline{j}R) = (\underline{i}N|\underline{j}R) + C_{Sj}(iN|RS)$
            Endif
        End i, j, N loop
    End R, S loop
    $(\underline{i}a|\underline{j}S) = \sum_N C_{Na}(\underline{i}N|\underline{j}S)$ (my $\underline{ij}$ pairs; all a, S, N)
    $(\underline{i}a|\underline{j}b) = \sum_S C_{Sb}(\underline{i}a|\underline{j}S)$ (my $\underline{ij}$ pairs; all a, b, S)
    $E_{\text{corr}}^{\text{MP2}} = E_{\text{corr}}^{\text{MP2}} + \sum_{ijab}(\underline{i}a|\underline{j}b)[2(\underline{i}a|\underline{j}b) - (\underline{i}b|\underline{j}a)]/(\epsilon_i + \epsilon_j - \epsilon_a - \epsilon_b)$ (my $\underline{ij}$ pairs; all a, b)
End I loop
Global summation of $E_{\text{corr}}^{\text{MP2}}$

FIGURE 9.2
The parallel direct MP2 algorithm P1 derived by making only small changes to the scalar algorithm in Figure 9.1. R, S, M, and N represent shells of atomic orbitals. Occupied orbitals, i, can be processed in batches, I, to reduce the storage requirement. Distributed indices are underlined.

of the M, N shell pairs is employed in the initial computation of the two-electron integrals and in the first quarter transformation; in the second quarter transformation and throughout the rest of the algorithm, the work is distributed over occupied orbital pairs i, j. This distribution of work necessitates two global communication steps. The first step is a global summation of the contributions to the quarter-transformed integrals, $(iN|RS)$, after the first quarter transformation. This summation is necessary because each process handles only a subset of the M, N pairs and therefore computes only a partial contribution to the quarter-transformed integrals. The second communication step is a global summation of the individual contributions to the correlation energy from each process. The algorithm distributes the integral arrays that are generated in full, namely, the half-, three-quarter-, and fully-transformed integrals, $(iN|jS)$, $(ia|jS)$, and $(ia|jb)$, respectively, and these arrays are distributed over occupied pairs i, j using the same distribution as for the work. By using the same distribution for work and data, communication can be minimized because each process handles only local data.

Comparison of Figures 9.1 and 9.2 shows that only a few, straightforward, modifications of the serial code are required to implement the parallel algorithm P1, and the specific changes are as follows. In the parallel version, a simple test is performed within the M, N loop to decide whether the current M, N pair is to be processed locally, and a similar test is performed on the i, j pairs in the second quarter transformation. Additionally, two communication steps have been added: a global summation after the first quarter transformation and a global summation of the correlation energy contributions at the end. Each of these summations is accomplished with a single function call to a global communication operation, in the first case doing an all-reduce operation (for example, `MPI_Allreduce`) and in the second case by means of an all-to-one reduce (for example, `MPI_Reduce`). In addition to these changes, the sizes of the integral arrays $(iN|jS)$, $(ia|jS)$, and $(ia|jb)$ have been modified in the parallel version because only a subset of the i, j pairs are stored and processed locally.

Let us develop a simple performance model for the algorithm P1. The only time-consuming communication step is the global summation of the quarter-transformed integrals. This summation is performed $\approx \frac{1}{2}n_I n_{\text{shell}}^2$ times, each summation involving integral arrays of length $\approx n_i n(n/n_{\text{shell}})^2$, where n_{shell}, n_I, and n_i denote the number of shells, the number of batches of occupied orbitals, and the number of occupied orbitals in a batch, respectively. Using $n_I n_i = o$, this yields a total number of $\approx \frac{1}{2}on^3$ integrals to be added in the global summation. Employing the performance model for Rabenseifner's all-reduce algorithm (Eq. 3.8), which is valid when the number of processes, p, is a power of two, the communication time can be expressed as follows

$$\begin{aligned} t_{\text{comm}}^{\text{P1}}(p) &= \frac{1}{2}n_I n_{\text{shell}}^2 \left[2\log_2 p\alpha + 2\frac{p-1}{p} n_i n \left(\frac{n}{n_{\text{shell}}}\right)^2 (\beta + \gamma/2) \right] \\ &= n_I n_{\text{shell}}^2 \log_2 p\alpha + \frac{p-1}{p} on^3(\beta + \gamma/2) \end{aligned} \tag{9.9}$$

employing the machine parameters α, β, and γ defined in chapter 5. The communication time is an increasing function of p, and this will adversely affect the parallel efficiency of the algorithm. Assuming that the computational time on one process, $t_{\text{comp}}(1)$, is known, and ignoring load imbalance, we can model the total execution time as

$$t_{\text{total}}^{\text{P1}}(p) = \frac{t_{\text{comp}}(1)}{p} + n_I n_{\text{shell}}^2 \log_2 p\alpha + \frac{p-1}{p} on^3(\beta + \gamma/2). \tag{9.10}$$

This yields the following expression for the speedup, $S(p) = t_{\text{total}}(1)/t_{\text{total}}(p)$

$$S^{\text{P1}}(p) = p \times \frac{t_{\text{comp}}(1)}{t_{\text{comp}}(1) + n_I n_{\text{shell}}^2 p \log_2 p\alpha + (p-1)on^3(\beta + \gamma/2)}. \tag{9.11}$$

The term multiplying p on the right hand side of Eq. 9.11 is a decreasing function of p, and the P1 algorithm is not strongly scalable. This lack of scalability is caused by the communication requirement, which increases with p and becomes a bottleneck for large process counts. The number of processes that can be efficiently utilized with this algorithm will depend on the machine parameters α, β, and γ, and also on the size of the molecule and the basis set employed. We will discuss the parallel performance of the P1 algorithm in more detail in section 9.5.

9.4 High-Performance Parallelization

To achieve a parallel MP2 implementation that is scalable, the simple global communication approach used in the P1 algorithm in the previous section must be abandoned. It is necessary, instead, to use a communication scheme in which the communication time per process decreases with increasing process counts, and in the following we will discuss a high-performance parallel direct MP2 algorithm, designated P2, using one-sided communication to achieve this. Two versions of the P2 algorithm have been implemented, using static load balancing and dynamic load balancing by means of a manager–worker scheme. We have included both implementations because parallel performance for large process counts can generally be improved by using a dynamic work distribution, although a manager–worker approach leads to reduced efficiency for small process counts because one process is dedicated to distributing work to the other processes.

An outline of the P2 algorithm is shown in Figure 9.3. Like the P1 algorithm of the previous section, P2 is based on the scalar algorithm of Figure 9.1, and both parallelizations have thus been realized without sacrificing scalar performance. In the P2 algorithm, the work distribution for computation of the two-electron integrals and the first and second quarter transformations is handled by distribution of the outer shell pairs R, S. The static version of the algorithm employs a round-robin distribution of the R, S pairs, whereas the dynamic version uses a dedicated manager process to distribute the R, S pairs. The manager keeps a list of R, S shell pairs, sorted according to decreasing size, and sends shell pairs to worker processes upon receiving requests for tasks. The manager–worker model used here is described in more detail in the discussion of dynamic load balancing in the two-electron integral transformation in chapter 7. Once an R, S shell pair has been allocated to a process, whether statically or dynamically, the shell quartet $(MN|RS)$ of two-electron integrals is computed for all M, N, and the first and

Loop over I batches

  While there are R, S shell pairs left

    Get my next $\underline{RS}$ pair ($S \leq R$)

    Loop over M, N shells ($N \leq M$)
      Compute $(MN|RS)$
      Loop over $i \in I$
        $(iN|\underline{RS}) = (iN|\underline{RS}) + C_{Mi}(MN|\underline{RS})$
        $(iM|\underline{RS}) = (iM|\underline{RS}) + C_{Ni}(MN|\underline{RS})$
      End i loop
    End M, N loop

    Loop over $i \in I$, all j
      Loop over all N
        $(iN|jS) = (iN|jS) + C_{Rj}(iN|\underline{RS})$
        $(iN|jR) = (iN|jR) + C_{Sj}(iN|\underline{RS})$
      End N loop
      Send $(iN|jS)$ and $(iN|jR)$ contributions
      to process holding current ij pair
    End i, j loop

  End while

  $(\underline{i}a|\underline{j}S) = \sum_N C_{Na}(\underline{i}N|\underline{j}S)$ (my $\underline{ij}$ pairs; all a, S, N)

  $(\underline{i}a|\underline{j}b) = \sum_S C_{Sb}(\underline{i}a|\underline{j}S)$ (my $\underline{ij}$ pairs; all a, b, S)

  $E^{\mathrm{MP2}}_{\mathrm{corr}} = E^{\mathrm{MP2}}_{\mathrm{corr}} + \sum_{ijab}(\underline{i}a|\underline{j}b)[2(\underline{i}a|\underline{j}b) - (\underline{i}b|\underline{j}a)]/(\epsilon_i + \epsilon_j - \epsilon_a - \epsilon_b)$
  (my $\underline{ij}$ pairs; all a, b)

End I loop

Global summation of $E^{\mathrm{MP2}}_{\mathrm{corr}}$

FIGURE 9.3
The high-performance parallel direct MP2 algorithm P2. R, S, M, and N denote shells of atomic orbitals. Occupied orbitals, i, can be processed in batches, I, to reduce the storage requirement. Distributed indices are underlined. R, S pairs are distributed either statically or dynamically as explained in the text.

second quarter transformations are then carried out for this set of integrals. In the second quarter transformation each process computes the contribution to the half-transformed integrals $(iN|jS)$ and $(iN|jR)$ from the locally held quarter-transformed integrals and for all i, j pairs. Because R, S pairs are distributed, individual processes will generate only a partial contribution to the half-transformed integrals, and these contributions will then be sent to the process that is to store the corresponding i, j pair and added into a local array of half-transformed integrals held by that process. To make the redistribution of the half-transformed integrals efficient, we use a one-sided message-passing scheme, which is implemented by means of multiple threads and uses separate threads for communication and computation (this scheme is discussed in more detail in section 4.4). After completion of the second quarter transformation, all processes hold a set of half-transformed integrals $(iN|jS)$ for all N, S and for a subset of the i, j pairs. The remainder of the transformation and the computation of the correlation energy are carried out by each process independently using only local data, processing the subset of the i, j pairs corresponding to the locally held half-transformed integrals. When all processes have finished computing their contribution to the correlation energy, a (fast) global summation is performed to add the individual contributions.

To develop a performance model for the P2 algorithm, we need to analyze the communication requirement in more detail. The only nonnegligible communication step in the algorithm is the sending of half-transformed integrals between processes during the integral transformation. This step involves, for each i, j pair, sending the contributions to the integrals $(iN|jS)$ and $(iN|jR)$ from the process where they were computed to the process that is to hold the current i, j pair. If the current i, j pair is to be held locally, the generated half-transformed integrals will simply be kept by the process that computed them without the need for any communication. If there are p processes, a fraction equal to $1/p$ of the contributions to the half-transformed integrals generated by a process will correspond to local i, j pairs, whereas the remaining fraction $(p-1)/p$, representing i, j pairs to be subsequently handled by other processes, must be sent to remote processes. Each process handles the fraction $1/p$ of the total work, and, consequently, a process must send the fraction $(p-1)/p^2$ of the total number of half-transformed integrals to remote processes where they will be added into local arrays. The sends are carried out inside a nested loop over batches I, the local R, S pairs, $i \in I$, and all j, yielding a total number of sends per process of $\approx 2n_I \frac{1}{2} n_{\text{shell}}^2 n_i o(p-1)/p^2 = n_{\text{shell}}^2 o^2 (p-1)/p^2$, where n_I and n_i are the number of I batches and the number of occupied orbitals in a batch, respectively. The length of each message is the number of integrals in the chunks of the $(iN|jS)$ or $(iN|jR)$ arrays that are sent between processes, and it is on average n^2/n_{shell}. Using the usual model, $t_{\text{send}} = \alpha + l\beta$, for sending a message of length l, the communication time for the P2 algorithm can be expressed as

$$t_{\text{comm}}^{\text{P2}}(p) = n_{\text{shell}}^2 o^2 \frac{p-1}{p^2} \left(\alpha + \frac{n^2}{n_{\text{shell}}} \beta \right). \tag{9.12}$$

The send operations employed are non-blocking and can be overlapped with computation to some extent. Our performance model will take this into account by using effective α and β values measured for overlapped communication and computation. From Eq. 9.12 it is clear that the overall communication requirement for the P2 algorithm, namely, the total amount of data that must be sent per process, $O(n_{\text{shell}} o^2 n^2/p)$, is roughly inversely proportional to the number of processes, which is a prerequisite for a strongly scalable algorithm. The total execution time for P2 can then be modeled as

$$t_{\text{total}}^{\text{P2}}(p) = \frac{t_{\text{comp}}(1)}{p} + n_{\text{shell}}^2 o^2 \frac{p-1}{p^2}\left(\alpha + \frac{n^2}{n_{\text{shell}}}\beta\right) \tag{9.13}$$

where $t_{\text{comp}}(1)$ is the single-process execution time, and the work is assumed to be evenly distributed. Using Eq. 9.13, we get the following expression for the speedup for P2

$$S^{\text{P2}}(p) = p \times \frac{t_{\text{comp}}(1)}{t_{\text{comp}}(1) + n_{\text{shell}}^2 o^2((p-1)/p)\left[\alpha + (n^2/n_{\text{shell}})\beta\right]}. \tag{9.14}$$

The speedup predicted by Eq. 9.14 is approximately proportional to p, indicating that the P2 algorithm is strongly scalable (as long as sufficient work exists to enable good load balance, that is, $o^2 \gg p$) and should be able to sustain a nearly constant efficiency as the number of processes increases. We have developed the performance model for P2 without specifying whether we consider the static or dynamic version because the same model is applicable in both cases, within limits. Equation 9.14 is valid for the static version, provided that load imbalance is negligible. As the number of processes grows large, however, the static version will experience load imbalance, and the speedups will be lower than predicted by the model. For the dynamic version of P2, load imbalance is not an issue even for very large numbers of processes, but the right hand side of Eq. 9.14 should be multiplied by $(p-1)/p$ because only $p-1$ processes are engaged in computation. This limits the maximum speedup to $p-1$ for the dynamic version. For large process counts where $p-1 \approx p$, this can be ignored, but for small numbers of processes ($p \lesssim 16$), the resulting loss of speedup and efficiency is nonnegligible. As demonstrated in chapter 7, the extra communication required to dynamically distribute the work is negligible and, therefore, can be omitted from Eq. 9.14.

In practice, we would use the static work distribution in P2 for $p \lesssim 16$, where load imbalance should be negligible in most cases, and use the dynamic distribution for larger process counts. Under those circumstances the performance model in Eq. 9.14 is valid and should give a realistic representation of the performance that can be obtained with the P2 algorithm. Although a manager–worker model has been employed here to distribute the work dynamically, it is also possible to implement dynamic load balancing without a dedicated manager process. This approach prevents the loss of efficiency incurred by not doing computation on the manager and has been explored in the context of parallel MP2 energy computation (see section 9.6). We will investigate the parallel performance of the P2 algorithm further in the next section.

9.5 Performance of the Parallel Algorithms

Let us analyze in more detail the parallel performance that can be obtained with the two parallel algorithms P1 and P2 described in the previous sections. The major features of the two algorithms are summarized in Table 9.1. P1 and P2 require the same amount of computation, and they can both utilize $O(o^2)$ processes, but their communication requirements differ. The amount of communication required for P1 is relatively small, $O(on^3)$, but the algorithm uses global communication, and the communication requirement per process does not decrease as the number of processes increases. For P2, which uses one-sided message-passing, the communication requirement, $O(n_{\text{shell}}o^2n^2/p)$, is inversely proportional to the number of processes. Consequently, the communication time for P2 should not become a bottleneck as the number of processes grows. This difference in communication patterns for P1 and P2 is the root of the performance differences obtained for the two algorithms as we will illustrate below. As mentioned earlier, the P1 and P2 algorithms can divide the occupied orbitals into batches and do the integral transformation for one batch at a time if the global memory available is insufficient for performing the integral transformation in one pass. This necessitates recomputation of the two-electron integrals, which must be evaluated for each batch. Thus, for cases where the number of required batches decreases with increasing process counts because more aggregate memory becomes available, superlinear speedups can be obtained for P1 and P2. In all examples considered in this chapter, only one batch is required in the integral transformation for any process count, and the ideal speedup curve, therefore, corresponds to linear speedups, $S(p) = p$; superlinear speedups for the P2 algorithm are illustrated in section 5.4.

To make quantitative predictions for the performance of P1 and P2, we need values for the parameters characterizing the molecule, namely n, n_{shell}, and o, as well as the machine-specific parameters α, β, and γ. We will consider the parallel performance of P1 and P2 on a Linux cluster[4] for the face-to-face isomer of the uracil dimer[5] ($C_8N_4O_4H_8$) employing the correlation-consistent double-ζ basis set cc-pVDZ and including all orbitals in the correlation procedure. For this system, the number of basis functions and shells are $n = 264$ and $n_{\text{shell}} = 104$, and the number of occupied orbitals o is 58. We first performed a number of benchmarks to determine the values for the machine parameters using the MPICH2[6] version of the Message-Passing Interface (MPI).* The values of the latency α and the inverse bandwidth β appropriate for communication that is overlapped with computation were found to be $\alpha = 9.9\ \mu s$ and $\beta = 15.7$ ns/word (= 1.96 ns/byte), and these values can be used in the performance model for P2. For P1, the machine parameters were determined

* We require that MPI be thread-safe, and this currently necessitates using the TCP/IP protocol running on IPoIB on our InfiniBand cluster. This results in lower performance than would be obtained with a native InfiniBand interface.

by fitting timings for the all-reduce operation to Eq. 3.8 (for process counts that are a power of two). We found the latency to be $\alpha = 130\ \mu s$, and the value for $\beta + \gamma/2$ was determined to be 58 ns/word (7.3 ns/byte). Note that in the performance models we use values of β and γ given in time units per word, using 8 byte words, because we have expressed the message length in terms of double precision words. Finally, for the performance models, the single-process computational time, $t_{\mathrm{comp}}(1)$, is also required, and $t_{\mathrm{comp}}(1)$ was found to be 2462 seconds. In this case, it was possible to perform the computation with one process (using only one pass in the integral transformation). In general, however, it may be necessary to obtain an effective floating point rate for the application and use this to estimate the single-process execution time, or, alternatively, to measure speedups relative to p_{min}, where p_{min} represents the smallest number of processes for which the computation can be performed. The compute units in the Linux cluster[4] used here for benchmarking are two-processor nodes, and to fully utilize the machine, two processes should be run per node. The performance results discussed here pertain to cases where only one process was run per node. Some performance degradation should be expected when using two processes per node because there will be some competition for memory and communication bandwidth. Full machine utilization can also be obtained by running two compute threads, instead of two processes, per node (a multi-threaded implementation of P2 is discussed in section 4.4).

Using the determined values for the various parameters, the performance models, Eqs. 9.10, 9.11, 9.13, and 9.14, yield the total execution times and speedups shown in Figures 9.4 and 9.5. The performance models predict P1

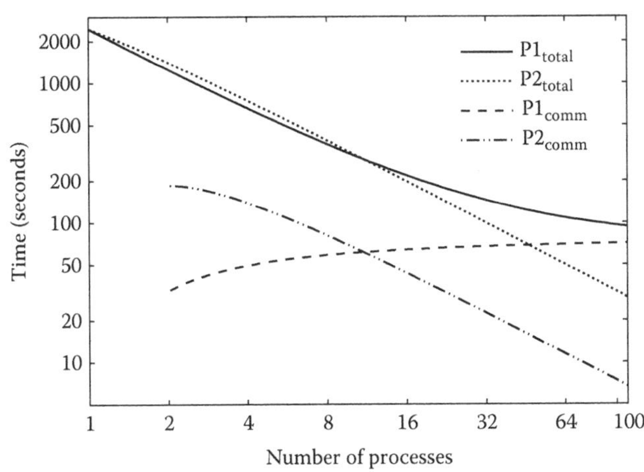

FIGURE 9.4
Log–log plot of predicted total execution times ($P1_{\mathrm{total}}$ and $P2_{\mathrm{total}}$) and communication times ($P1_{\mathrm{comm}}$ and $P2_{\mathrm{comm}}$) on a Linux cluster for the uracil dimer using the cc-pVDZ basis set. Ideal scaling corresponds to a straight line of negative unit slope.

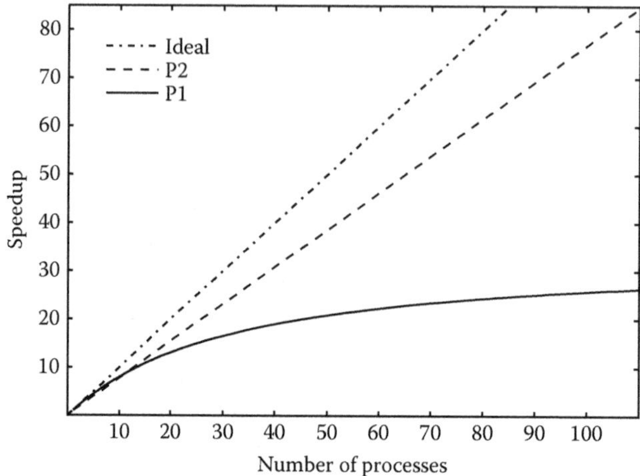

FIGURE 9.5
Speedups (relative to one process) predicted by the performance models for algorithms P1 and P2 on a Linux cluster for the uracil dimer using the cc-pVDZ basis set.

to be the fastest algorithm for small process counts and P2 to become faster than P1 as the number of processes increases. It is also clear from Figure 9.4 that the fraction of the total execution time spent doing communication increases rapidly with the number of processes for P1, whereas the communication time for P2 constitutes a nearly constant fraction of the total execution time as the number of processes increases. The speedups predicted by the performance models (Figure 9.5) illustrate the poor scalability of P1, caused by the communication overhead, and the maximum speedup that can be attained with P1 for the example considered is only about 30. The P2 algorithm, on the other hand, displays strong scalability as evidenced by the nearly linear speedup curve. Although the predicted efficiency for P2 is high, around 75%, it is significantly below 100% because the communication overhead, though scalable, is not negligible and constitutes approximately 25% of the total execution time.

To achieve high parallel efficiency with the P2 algorithm, the communication network must be capable of delivering a high bandwidth. Using a network with a significantly smaller bandwidth than in the example above will yield a lower parallel efficiency: if the InfiniBand network used in the examples above is replaced by a Gigabit Ethernet network, the latency and inverse bandwidth for communication overlapped with computation will be around 11 μs and 68 ns/word, respectively; although the latency is nearly the same as for the InfiniBand network in this particular case, the bandwidth delivered by the Ethernet network is more than 4 times lower than that observed for InfiniBand. The consequence of the lower bandwidth is illustrated in Figure 9.6, which shows the speedups predicted by the performance model for P2 using the network parameters for both InfiniBand and Gigabit Ethernet.

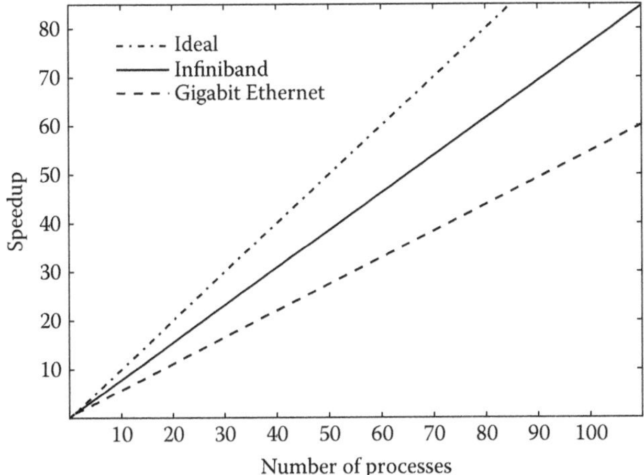

FIGURE 9.6
Predicted speedups (Eq. 9.14) for the P2 algorithm using network performance parameters for InfiniBand and Gigabit Ethernet (using the same test case as in Figure 9.5).

The difference in speedup curves for these two cases demonstrates that network performance characteristics can have significant impact on the parallel performance of an algorithm.

The measured total execution times and communication times for P1 and P2 on the employed Linux cluster[4] are illustrated in Figure 9.7. For larger

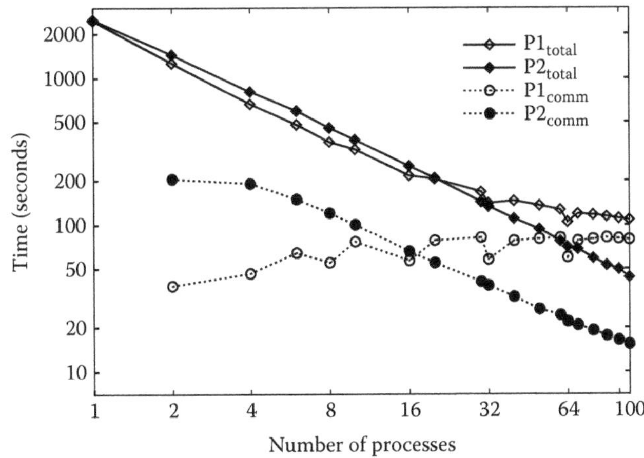

FIGURE 9.7
Log–log plot of measured total execution times ($P1_{total}$ and $P2_{total}$) and communication times ($P1_{comm}$ and $P2_{comm}$) on a Linux cluster[4] for the uracil dimer using the cc-pVDZ basis set. Only the static version of the P2 algorithm is shown. Ideal scaling corresponds to a straight line of negative unit slope. See text for details.

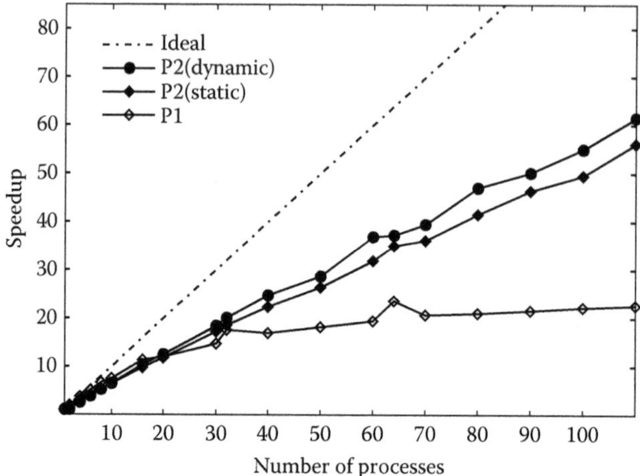

FIGURE 9.8
Measured speedups (relative to single-process timings) for the P1 and P2 algorithms on a Linux cluster[4] for the uracil dimer using the cc-pVDZ basis set.

process counts, P2 is significantly faster than P1 (about two and half times faster for $p = 110$), but in the interval $2 < p \lesssim 16$, P1 is the fastest algorithm. For both algorithms, the measured performance agrees well with the predictions of the performance model (Figure 9.4): the performance models correctly predict the trends in the total execution times and communication times for P1 and P2, and the models are adequate for doing performance analyses of the algorithms. Dips in the timings, and corresponding increases in the speedups, are observed for P1 for process counts that are powers of two because the employed all-reduce algorithm in P1 in that case is more efficient. In the performance model for P1, we modeled the communication time using Eq. 9.9, which is valid when the number of processes is a power of two, and the predictions of the performance model for P1 are therefore more accurate in those cases. The measured speedups for P1 and P2 are depicted in Figure 9.8, and the associated efficiencies are shown in Figure 9.9. For comparison, we have included measured speedups and efficiencies for both the static and dynamic versions of P2. Looking at the efficiencies in Figure 9.9, we note that, for small process counts, the P1 algorithm achieves a high parallel efficiency. The efficiency loss displayed by the dynamic version of P2 when the number of processes is small corresponds roughly to a fraction of $1/p$, as expected, and the efficiency therefore increases with the number of processes in a limited region. When p is small, the static version of P2 displays better performance than the dynamic version, but for larger process counts, where load imbalance becomes more pronounced, the dynamic version is faster. Finally, we note, that as the number of processes grows large, the P2

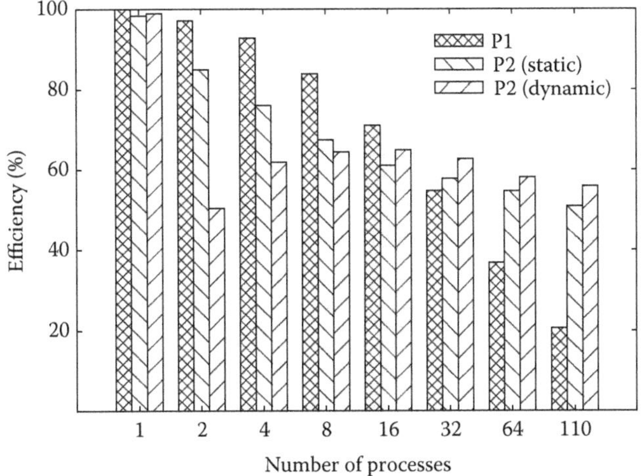

FIGURE 9.9
Measured efficiencies for the P1 and P2 algorithms on a Linux cluster[4] for the uracil dimer using the cc-pVDZ basis set.

algorithm is capable of maintaining a nearly constant efficiency, even though the problem size is rather small.

In summary, we have analyzed the performance of two parallel direct MP2 implementations derived from the same scalar algorithm: a simple parallel algorithm (P1), whose implementation requires minimal modifications to the scalar code, and a more sophisticated parallel implementation (P2) designed to be scalable. We have demonstrated that the simple parallelization strategy used in P1 yields excellent performance for small process counts ($p \lesssim 16$), and that the required global communication operation causes very poor performance when the number of processes is large, severely limiting the scalability of P1. We have also shown that it is possible to develop a scalable MP2 algorithm (P2) by using a communication scheme involving one-sided communication. Although the P2 algorithm is capable of providing sustained, high parallel efficiency as the number of processes increases, achieving high parallel efficiency is contingent upon having a communication network with a high bandwidth. Note that another high-performance parallel direct MP2 algorithm, which can exploit only one of the three index permutation symmetries in the integral computation but has a smaller communication requirement, is discussed in chapter 10 in the context of local MP2 theory. Both the P1 and P2 algorithms can display superlinear speedups for cases where the number of required batches of occupied orbitals decreases with increasing process counts as more global memory becomes available. In the discussion of speedups in section 5.4, superlinear speedup curves, as well as other inflated speedups computed relative to p-process timings, where $p > 1$, are illustrated for P1 and P2.

9.6 Further Reading

For an in-depth discussion of the MP2 method see, for example, Szabo and Ostlund[1] and Helgaker et al.[7] Many parallel MP2 implementations have been reported in the literature, and the algorithms mentioned below represent just a small selection of these, illustrating different parallelization strategies. Wong et al.[8] developed a parallel direct MP2 algorithm employing one-sided communication and using a dynamic distribution of shell pairs in the first half of the integral transformation and a static distribution of pairs of occupied orbitals in the remaining part. A somewhat similar parallelization strategy has been used by Schütz and Lindh.[9] These algorithms distribute data using the Global Array toolkit[10] and also employ a global array of computational tasks from which processes fetch tasks to be processed, thus obviating the need for a dedicated manager process. A distributed data parallel MP2 implementation has also been achieved using the Distributed Data Interface,[11] storing the large data arrays in the aggregate memory of a distributed memory computer. A parallelization strategy similar to P2, using static load balancing, has been employed in a massively parallel direct algorithm for computation of MP2 energies and gradients.[12] Parallel MP2 algorithms using disk I/O to store the half-transformed integrals have been developed as well, using a combination of dynamic distribution of two shell indices[13] or one shell index[14] followed by a static distribution of occupied orbital pairs.

References

1. Szabo, A., and N. S. Ostlund. *Modern Quantum Chemistry*, 1st revised edition, chapter 6. New York: McGraw-Hill, 1989.
2. Møller, C., and M. S. Plesset. Note on an approximation treatment for many-electron systems. *Phys. Rev.* 46:618–622, 1934.
3. Head-Gordon, M., J. A. Pople, and M. J. Frisch. MP2 energy evaluation by direct methods. *Chem. Phys. Lett.* 153:503–506, 1988.
4. A Linux® cluster consisting of nodes with two single-core 3.6 GHz Intel® Xeon® processors (each with 2 MiB of L2 cache) connected with a 4x Single Data Rate InfiniBand™ network using Mellanox Technologies MT25208 InfiniHost™ III Ex host adaptors and a Topspin 540 switch. The InfiniBand host adaptors were resident in a PCI Express 4x slot, reducing actual performance somewhat compared to using them in an 8x slot. MPICH2 1.0.5p4 was used with TCP/IP over InfiniBand IPoIB because this configuration provided the thread-safety required for some applications.
5. Leininger, M. L., I. M. B. Nielsen, M. E. Colvin, and C. L. Janssen. Accurate structures and binding energies for stacked uracil dimers. *J. Phys. Chem. A* 106:3850–3854, 2002.

6. MPICH2 is an implementation of the Message-Passing Interface (MPI). It is available on the World Wide Web at http://www.mcs.anl.gov/mpi/mpich2.
7. Helgaker, T., P. Jørgensen, and J. Olsen. *Molecular Electronic-Structure Theory*, chapter 14. Chichester, UK: John Wiley & Sons, 2000.
8. Wong, A. T., R. J. Harrison, and A. P. Rendell. Parallel direct four-index transformations. *Theor. Chim. Acta* 93:317–331, 1996.
9. Schütz, M., and R. Lindh. An integral-direct, distributed-data, parallel MP2 algorithm. *Theor. Chim. Acta* 95:13–34, 1997.
10. Nieplocha, J., R. J. Harrison, and R. J. Littlefield. Global arrays: A portable "shared-memory" programming model for distributed memory computers. In *Proceedings of the 1994 Conference on Supercomputing*, pp. 340–349. Los Alamitos: IEEE Computer Society Press, 1994.
11. Fletcher, G. D., M. W. Schmidt, B. M. Bode, and M. S. Gordon. The Distributed Data Interface in GAMESS. *Comp. Phys. Comm.* 128:190–200, 2000.
12. Nielsen, I. M. B. A new direct MP2 gradient algorithm with implementation on a massively parallel computer. *Chem. Phys. Lett.* 255:210–216, 1996.
13. Baker, J., and P. Pulay. An efficient parallel algorithm for the calculation of canonical MP2 energies. *J. Comp. Chem.* 23:1150–1156, 2002.
14. Ishimura, K., P. Pulay, and S. Nagase. A new parallel algorithm of MP2 energy evaluation. *J. Comp. Chem.* 27:407–413, 2006.

10

Local Møller–Plesset Perturbation Theory

The computational cost of most correlated electronic structure methods scales as a high-degree polynomial in the molecular size, typically $O(N^5)$–$O(N^7)$, where N is the size of the molecule. This scaling poses a serious challenge to the application of high-level quantum chemical methods to larger molecular systems, and to extend the scope of such methods, it is necessary to employ alternative formulations that reduce the computational complexity. Reduced-scaling approaches that exploit the inherently local nature of dynamic electron correlation have been developed for a number of correlated electronic structure methods, including second-order Møller–Plesset perturbation (MP2) theory. One such method, the local MP2 method LMP2, [1,2] employs an orbital-invariant formulation that makes it possible to express the energy in a basis of noncanonical orbitals. The delocalized nature of the canonical molecular orbitals used in conventional MP2 theory gives rise to the high-order polynomial scaling of the cost, and by permitting the use of localized orbitals, the local MP2 method offers the potential for a significantly reduced computational cost.

The application of local, reduced-scaling quantum chemical methods, nonetheless, places heavy demands on the computational resources, and to take full advantage of these methods it is necessary to develop algorithms tailored to application to massively parallel computers. Parallel algorithm development for local correlation methods entails special challenges because of the irregular data structures involved, making it more difficult to achieve an even distribution of work and data. In this chapter we will discuss massively parallel computation of energies with the LMP2 method, focusing on issues pertaining to parallelization of local correlation methods. We will first give a brief introduction to LMP2 theory, and we will then present a massively parallel LMP2 algorithm and discuss its parallel performance.

10.1 The LMP2 Equations

In LMP2 theory, the MP2 equations are expressed using an orbital-invariant formulation employing noncanonical orbitals, and a number of approximations are introduced to achieve reduced scaling of the computational cost.

The employed set of noncanonical orbitals are local in nature: the occupied orbital space is represented by localized occupied molecular orbitals, and the virtual (unoccupied) orbital space consists of atomic orbitals (AOs) projected into the virtual orbital space to ensure the required orthogonality of the occupied and unoccupied spaces.

In the orbital-invariant formulation, the closed-shell MP2 correlation energy can be expressed as follows[3]

$$E_{\mathrm{MP2}}^{\mathrm{corr}} = \sum_{ij} \mathrm{Tr}[\mathbf{K}_{ij}(2\mathbf{T}_{ji} - \mathbf{T}_{ij})] = \sum_{ijab} K_{ij}^{ab}\left(2T_{ij}^{ab} - T_{ij}^{ba}\right). \qquad (10.1)$$

The elements of the matrix $\mathbf{K}_{ij}$ are the two-electron integrals, $K_{ij}^{ab} = (ia|jb)$, and the elements of the matrix $\mathbf{T}_{ij}$ are the double-substitution amplitudes T_{ij}^{ab}. Indices i, j and a, b represent occupied and unoccupied orbitals, respectively.

The two-electron integrals $(ia|jb)$ are computed by transformation of the AO integrals $(\mu\nu|\lambda\sigma)$ in a four-step transformation similar to that used in conventional MP2 theory (cf. section 9.1)

$$(i\nu|\lambda\sigma) = \sum_{\mu} L_{\mu i}(\mu\nu|\lambda\sigma) \qquad (10.2)$$

$$(i\nu|j\sigma) = \sum_{\lambda} L_{\lambda j}(i\nu|\lambda\sigma) \qquad (10.3)$$

$$(ia|j\sigma) = \sum_{\nu} P_{\nu a}(i\nu|j\sigma) \qquad (10.4)$$

$$(ia|jb) = \sum_{\sigma} P_{\sigma b}(ia|j\sigma) \qquad (10.5)$$

using transformation matrices **L** and **P** instead of the molecular orbital coefficient matrix used in conventional MP2 theory. The matrix **L** transforms from the atomic orbital basis into the basis of localized occupied orbitals, and **P** is the projection matrix that transforms into the projected atomic orbital basis.

The double-substitution amplitudes from Eq. 10.1 are determined from a set of linear equations

$$\mathbf{R}_{ij} = \mathbf{K}_{ij} + \mathbf{F}\mathbf{T}_{ij}\mathbf{S} + \mathbf{S}\mathbf{T}_{ij}\mathbf{F} - \mathbf{S}\sum_{k}[F_{ik}\mathbf{T}_{kj} + F_{kj}\mathbf{T}_{ik}]\mathbf{S} = 0 \qquad (10.6)$$

which in LMP2 theory (unlike in conventional MP2 theory) must be determined by an iterative procedure. In this equation, $\mathbf{R}_{ij}$ is the residual matrix with elements R_{ij}^{ab}, **F** and **S** represent the virtual–virtual blocks of the Fock matrix and the overlap matrix, respectively, and F_{ik} is an element of the Fock matrix in the localized occupied orbital basis. The iterative solution of the residual equation, Eq. 10.6, is described in detail elsewhere.[4,5] It involves, in each iteration, computing the residual matrices $\mathbf{R}_{ij}$ from the current amplitudes T_{ij}^{ab}, using the computed residual elements to form the updates ΔT_{ij}^{ab} to the amplitudes, and then using the updated amplitudes in the next iteration.

Reduced computational scaling is achieved in LMP2 theory by neglecting the contribution to the correlation energy arising from interactions of orbitals that are spatially distant and by employing prescreening techniques throughout the integral transformation. Orbital interactions are neglected by including only double substitutions involving excitations out of pairs of spatially close localized occupied orbitals and into pairs of unoccupied orbitals (projected atomic orbitals) belonging to domains associated with the involved occupied orbitals; the domains are predetermined, and there is one domain for each occupied orbital, containing the atomic orbitals that contribute the most to the Mulliken population of the occupied orbital.

10.2 A Scalar LMP2 Algorithm

The dominant steps in the computation of the LMP2 energy are the integral transformation and the computation of the residual in the iterative procedure. Of these two steps, which account for the vast majority of the computational time, the integral transformation generally is by far the most time-consuming, although the time required to compute the residual (which must be done in each iteration) is nonnegligible.

In Figure 10.1 we show an outline of a scalar algorithm for computing the LMP2 energy.[5] Initially, the localized occupied molecular orbitals are formed. The orbital domains are then created, and the screening quantities that will be used in the integral transformation are computed. The two-electron integral transformation is then performed, generating the integrals that are subsequently needed in the iterative solution of the residual equation (Eq. 10.6) and

```
Localize occupied molecular orbitals
Create domains
Compute screening quantities required for integral transformation
Perform integral transformation generating (ai|bj)
Begin LMP2 iterations
    Compute LMP2 residual
    Compute ΔT
    Update T (use DIIS extrapolation if desired)
    Compute E_MP2^corr, ΔE_MP2^corr
    Check for convergence on ΔE_MP2^corr and ΔT
End LMP2 iterations
```

FIGURE 10.1
Outline of a scalar local MP2 algorithm.

the computation of the energy. In the iterative procedure, the LMP2 residual is first computed and then used to form the update to the double-substitution amplitudes, ΔT_{ij}^{ab}. Using ΔT_{ij}^{ab} and employing the DIIS (direct inversion in the iterative subspace) procedure, the double-substitution amplitudes **T** are then updated, and a new value for the LMP2 energy is computed. A check for convergence of the amplitudes and the energy is finally performed, and, if convergence has not been reached, a new iteration is started.

The screening procedures and the systematic neglect of certain orbital interactions employed in local correlation methods produce sparse data structures (for example, the two-electron integrals), and this sparsity must be utilized to achieve reduced computational scaling. Implementation of local correlation methods therefore requires an efficient way to handle such sparse data structures, storing and manipulating only the nonzero elements. Sparse matrices and higher-dimensional arrays can be handled by dividing the data structures into smaller, independent, dense data structures that are stored and manipulated individually. Alternatively, sparse data representations can be employed, which store the nonzero elements in contiguous memory locations (while keeping track of where the data fit into the full matrix or array) and also provide support for performing operations on the data structure as a single unit. The LMP2 implementation discussed in this chapter employs the latter approach, using a sparse data representation that supports the handling of sparse matrices and multidimensional arrays.[5]

10.3 Parallel LMP2

For an efficient parallel LMP2 implementation we will require distribution of all large data arrays to avoid storage bottlenecks, parallelization of all computationally significant steps, and a communication scheme that does not seriously impair parallel performance as the number of processes increases. To ensure portability of the code, we will try to use a communication scheme involving only collective communication. In chapter 9 we investigated an MP2 algorithm (designated P2) using one-sided communication, implemented by means of separate communication threads, that yielded a scalable algorithm with very good parallel performance. This communication scheme, however, required using a thread-safe version of the Message-Passing Interface (MPI), reducing portability of the code.

The various sparse arrays, including the two-electron integrals and the double-substitution amplitudes, will be handled by a sparse data representation that also provides support for parallel operations such as converting a replicated array to a distributed array and redistributing an array using a different distribution scheme;[5] a set of generalized contraction routines developed to perform contractions of these distributed sparse multidimensional arrays will be employed as well.

In the following we will use a matrix notation for the double-substitution amplitudes, $\mathbf{T}_{ij}$, two-electron integrals, $\mathbf{K}_{ij}$, and residual elements, $\mathbf{R}_{ij}$, where, for instance, $\mathbf{T}_{ij}$ contains all the double-substitution amplitudes T_{ij}^{ab} for a fixed ij pair. Using the matrix notation is convenient in the discussion but does not reflect the actual data representation. The double-substitution amplitudes, integrals, and residual elements are stored as four-dimensional sparse arrays using the employed sparse data representation. In this representation, however, indices can be frozen: for instance, freezing the indices i and j in the four-index array $\mathbf{T}$ with indices i, j, a, and b allows the sub-array $\mathbf{T}_{ij}$ to be manipulated as a matrix.

In the following we will discuss parallel implementation of the two dominant computational steps in the LMP2 procedure, namely the two-electron integral transformation and the computation of the residual. Parallelization of other computationally less demanding, but nonnegligible, steps is straightforward.[5]

10.3.1 Two-Electron Integral Transformation

Apart from involving sparse data structures, a more complicated screening protocol, and different transformation matrices, the LMP2 integral transformation is similar to that of a conventional MP2 approach. We will therefore explore how a parallel algorithm that has been used successfully in a parallel conventional MP2 algorithm[6] will fare for the LMP2 case. In addition to making the changes appropriate for the LMP2 method, we will implement the algorithm using only collective communication to increase portability. The AO integrals as well as the partially and fully transformed integrals will be handled using the parallel sparse data representation mentioned above.

The parallel two-electron integral transformation algorithm, as employed here, is outlined in Figure 10.2. The algorithm computes the AO integrals and performs four separate quarter transformations, yielding the fully transformed two-electron integrals. One of the three index permutation symmetries of the integrals is used, resulting in a four-fold redundancy in the integral computation. Work and data are distributed by distribution of pairs of shells of atomic orbitals M, N in the first part of the algorithm and by distribution of pairs of localized occupied orbitals i, j in the last part of the transformation. Schwarz screening, as well as other screening using the transformation matrices $\mathbf{P}$ and $\mathbf{L}$, is performed after the distribution of tasks (possibly eliminating more work for some processes than for others). To switch from the M, N to the i, j distribution, a communication step is required to redistribute the half-transformed integrals. This redistribution takes place when all of the half-transformed integrals have been generated, and it is the only communication step in the integral transformation. It can be accomplished by an all-to-all scatter operation in which each process scatters its integrals, in a prescribed way, to all the other processes. If the number of half-transformed integrals is designated n_{half} and the number of processes is p, there will be n_{half}/p integrals per process; using an all-to-all scatter algorithm that has a

While get_pair(M, N) (get next available MN pair)
  Allocate all AO integral blocks ($\underline{M}R|\underline{N}S$) for current $\underline{M}, \underline{N}$ pair
  Compute ($\underline{M}R|\underline{N}S$)
  Allocate all ($\underline{M}R|\underline{N}j$) blocks for current $\underline{M}, \underline{N}$ pair
  1st quarter transf.: $(\underline{M}R|\underline{N}j) = \sum_S(\underline{M}R|\underline{N}S)L(S, j)$
  Allocate all ($\underline{M}i|\underline{N}j$) blocks for current $\underline{M}, \underline{N}$ pair
  2nd quarter transf.: $(\underline{M}i|\underline{N}j) = \sum_R(\underline{M}R|\underline{N}j)L(R, i)$
End while

Redistribute half-transformed integrals: $(\underline{M}i|\underline{N}j) \rightarrow (M\underline{i}|N\underline{j})$

Allocate all ($M\underline{i}|b\underline{j}$) blocks for local i, j pairs
3rd quarter transf.: $(M\underline{i}|b\underline{j}) = \sum_N(M\underline{i}|N\underline{j})P(N, b)$
Allocate all ($a\underline{i}|b\underline{j}$) blocks for local i, j pairs
4th quarter transf.: $(a\underline{i}|b\underline{j}) = \sum_M(M\underline{i}|b\underline{j})P(M, a)$

FIGURE 10.2
Outline of the parallel two-electron integral transformation used in the LMP2 algorithm. Indices M, R, N, S represent shells of atomic orbitals; i, j are localized occupied molecular orbitals; and a, b denote projected atomic orbitals. Distributed indices are underlined. The algorithm utilizes the M, N and i, j permutational symmetries. All the integral allocation steps are preceded by integral screening (not shown). Indices M, N are distributed throughout the first half of the transformation, while the remainder of the transformation uses distributed i, j indices. The only communication step required is the redistribution of the half-transformed integrals to switch from the M, N to the i, j distribution.

small bandwidth term and is expected to perform well for long messages,[7] the total time required to redistribute the integrals can be modeled as

$$t_{\rm comm} = (p-1)\alpha + \frac{n_{\rm half}}{p}\beta \tag{10.7}$$

where α and β represent the latency and inverse bandwidth, respectively. The bandwidth term is inversely proportional to p, and if the message length is sufficiently long to make the latency term negligible (a reasonable assumption because $n_{\rm half}$ tends to be very large), the redistribution of the integrals is a strongly scalable step.

For canonical MP2 theory, the above integral transformation has been shown to yield high parallel efficiency[6] when the integrals are redistributed by means of one-sided message-passing, which allows overlapping communication with computation to achieve a higher effective bandwidth. Because we use a collective communication step to redistribute the integrals in the LMP2 algorithm, this overlap of communication and computation cannot be achieved. The number of integrals to be redistributed in the LMP2 case, however, is much smaller than for canonical MP2; the number of half-transformed integrals to redistribute is only $O(n)$ as opposed to $O(o^2n^2)$ in canonical MP2 theory, where n is the number of basis functions and o is the number of occupied orbitals. As we shall see in section 10.3.3 this makes the communication overhead for the LMP2 integral transformation negligible, even though the computational complexity of the LMP2 procedure is also much smaller than that of the conventional MP2 method. Direct implementations of canonical MP2 theory often perform the integral transformation in several batches, thus reducing the overall memory requirement but increasing the computational cost because the AO integrals must be recomputed in each batch. The LMP2 algorithm, however, completes the entire integral transformation in one pass because the reduced storage requirement of the LMP2 approach relative to conventional MP2 makes integral storage unlikely to constitute a bottleneck.

10.3.2 Computation of the Residual

The computation of the residual is the dominant step in the iterative procedure. From Eq. 10.6, we see that a given residual matrix $\mathbf{R}_{ij}$, with elements R_{ij}^{ab}, contains contributions from the integrals and double-substitution amplitudes with the same occupied indices, $\mathbf{K}_{ij}$ and $\mathbf{T}_{ij}$, respectively, as well as from the double-substitution amplitudes $\mathbf{T}_{ik}$ and $\mathbf{T}_{kj}$. The contributions from $\mathbf{T}_{ik}$ and $\mathbf{T}_{kj}$ complicate the efficient parallelization of the computation of the residual and make communication necessary in the iterative procedure. The double-substitution amplitudes can either be replicated, in which case a collective communication (all-to-all broadcast) step is required in each iteration to copy the new amplitudes to all processes; or the amplitudes can be distributed, and each process must then request amplitudes from other processes as needed throughout the computation of the residual. Achieving high parallel efficiency in the latter case requires the use of one-sided message-passing.

In keeping with our intent to use only readily portable types of communication, we will here replicate the amplitudes and perform the collective communication operations that this approach entails. The parallel computation of the residual then proceeds as shown in Figure 10.3. Each process computes the residual matrices $\mathbf{R}_{ij}$ for the ij pairs for which the integral matrices $\mathbf{K}_{ij}$ are stored locally; to do so, only locally stored data are needed because the required $\mathbf{K}_{ij}$ integrals as well as all of the double-substitution amplitudes are available locally. After computing a residual matrix $\mathbf{R}_{ij}$, a process will compute the corresponding double-substitution amplitude update $\Delta\mathbf{T}_{ij}$ and use this update to form the new double-substitution amplitudes $\mathbf{T}_{ij}$. When all processes have computed the new $\mathbf{T}_{ij}$ amplitudes for their ij pairs, a collective operation is performed to put a copy of the entire $\mathbf{T}$ array on all processes. This is accomplished by an all-to-all broadcast operation in which each process broadcasts the locally computed $\mathbf{T}_{ij}$ matrices to every other process. The cost of performing this communication step can be estimated as follows. Let the number of processes and the total number of double-substitution amplitudes be p and n_{ampl}, respectively (that is, there are n_{ampl}/p amplitudes

```
While get_local_pair(i, j)        (get next local ij pair)
    Initialize R_ij to K_ij
    Update R_ij with FT_ijS and ST_ijF terms
    Update R_ij with S Σ_k[F_ik T_kj + F_kj T_ik]S term
    Compute ΔT_ij from R_ij
    Compute new T_ij
End while
All-to-all broadcast of local T_ij matrices
```

FIGURE 10.3
Outline of parallel computation of the LMP2 residual, which is required in each iteration. The update of the double-substitution amplitudes is shown as well. Every process loops over the local ij pairs, defined as those ij pairs for which the transformed two-electron integrals $\mathbf{K}_{ij}$ reside locally. In the final communication step, each process broadcasts all of its $\mathbf{T}_{ij}$ matrices to every other process, giving all processes a copy of the entire updated $\mathbf{T}$ array.

per process); using a performance model for a recursive doubling all-to-all broadcast (Eq. 3.4), the communication time can then be modeled as

$$t_{\text{comm}} = \log_2 p\alpha + n_{\text{ampl}}\frac{p-1}{p}\beta. \tag{10.8}$$

As opposed to the collective communication operation used in the integral transformation, this communication step is not scalable; the communication time will increase with p, or, if the latency term can be neglected, remain nearly constant as p increases. This communication step is therefore a potential bottleneck, which may cause degrading parallel performance for the LMP2 procedure as the number of processes increases. To what extent this will happen depends on the actual time required for this step compared with the other, more scalable, steps of the LMP2 procedure, and we will discuss this issue in more detail in the following section.

10.3.3 Parallel Performance

Let us investigate the parallel performance of the LMP2 algorithm described above. We will first look at the performance for the test case that was also used in chapter 9 for the parallel canonical MP2 algorithms (Figure 9.8), namely the uracil dimer, using a cc-pVDZ basis set and running on a Linux cluster.[8] The speedups measured for the parallel LMP2 algorithm for this case are shown in Figure 10.4. For this relatively small molecule, the LMP2 transformation is only slightly faster (about 2%) than the canonical MP2 transformation; the computational savings obtained by restricting the correlation space according to the local correlation model are roughly offset by the additional work required as a consequence of using a larger dimension for the virtual space (the number of virtual orbitals used in the LMP2 method equals the number of basis functions). The LMP2 integral transformation yields a nearly linear

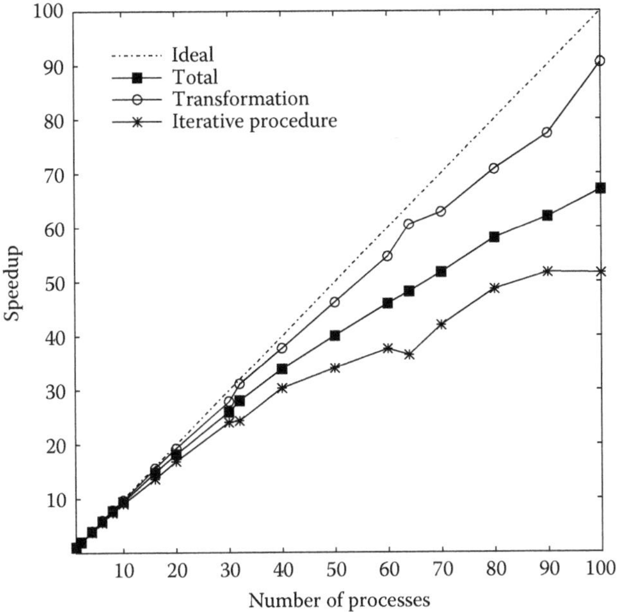

FIGURE 10.4
Measured parallel LMP2 speedups for the uracil dimer using the cc-pVDZ basis set. Timings were obtained on a Linux cluster;[8] speedups were computed (from measured wall times) relative to the execution time required using one process.

speedup curve and achieves a speedup of 90 when running on 100 processes. The redistribution of the integrals is scalable, as predicted by the model in Eq. 10.7, and it takes up about 2% of the total execution time for the integral transformation for all process counts investigated. Load imbalance in the integral transformation, though not severe, is the main reason for the slight deterioration of the performance observed for large process counts. High parallel efficiency can be achieved using a collective communication step to redistribute the integrals in the LMP2 transformation, and using the integral transformation from above in a conventional MP2 algorithm in conjunction with an efficient one-sided message-passing scheme also yields very good parallel performance.[6] The communication requirement of $O(o^2n^2)$ for conventional MP2, moreover, is smaller than that of the high-performance MP2 algorithm P2 from section 9.4, but the tradeoff for the reduced communication requirement is a four-fold redundancy in the integral computation, compared with only a two-fold redundancy for the P2 algorithm.

Looking at the parallel performance for the iterative procedure, the speedups are significantly lower than for the integral transformation, and for the uracil dimer the iterative procedure achieves a speedup of 52 for 100 processes. The nonscalable collective communication step required in each iteration is the primary factor contributing to lowering the speedup, but a

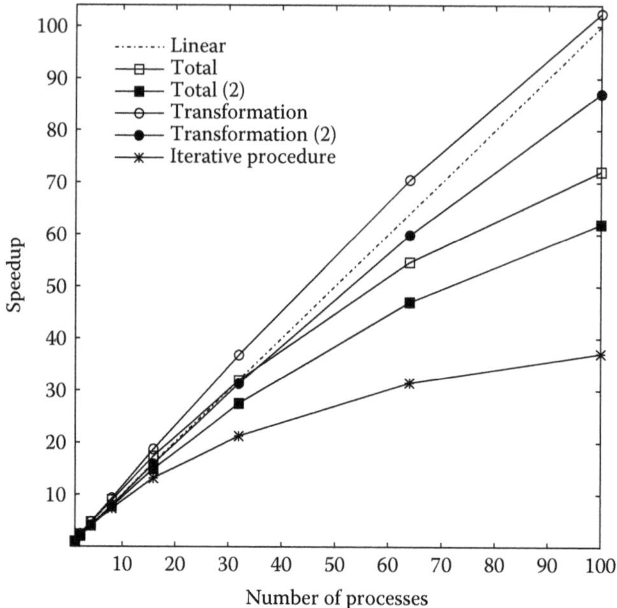

FIGURE 10.5
Measured speedups for the parallel LMP2 procedure for the linear alkane $C_{32}H_{66}$ using the cc-pVDZ basis set. Timings (wall times) were measured on a Linux cluster.[8] Speedups labeled with (2) are computed relative to two-process timings, and other speedups are measured relative to single-process timings. Paging in the integral transformation when using a single process artificially elevates the speedups measured relative to single-process timings, and speedups for the transformation and the overall procedure are therefore also shown measured relative to two-process timings, setting the speedup $S(p = 2)$ equal to 2.

small amount of load imbalance for large process counts contributes as well. The iterative procedure takes up around 30% of the total LMP2 execution time for this case (which required 13 iterations to converge the energy to 10^{-7} hartrees), so the parallel efficiency for this step is significant with respect to the overall performance for the LMP2 procedure. The speedups for the entire LMP2 procedure yield an efficiency of 67% when running on 100 processes; the speedups are nearly ideal up to around 32 processes after which the performance starts to decay.

Let us also examine the parallel performance for a larger, linear molecule for which the LMP2 procedure yields significant computational savings. Figure 10.5 shows speedups for the LMP2 procedure measured for the linear alkane $C_{32}H_{66}$ with the cc-pVDZ basis. Speedups for the entire LMP2 procedure and for the integral transformation are very similar to those obtained for the uracil dimer, although a bit lower: when using 100 processes, the speedups are 62 and 87 for the overall procedure and the integral transformation, respectively. The step whose parallel performance is most affected by the increased sparsity is the iterative procedure, which yields an efficiency of

37% on 100 processes. When sparsity can be exploited to a large degree, however, the iterative procedure becomes a less significant computational step, requiring only about 12% of the total execution time for 100 processes and less for smaller process counts. The less favorable scaling of the iterative procedure, therefore, does not have a large impact on the overall parallel efficiency for the LMP2 procedure. The performance obtained for this linear alkane, for which sparsity is utilized to a large extent, demonstrates that the integral screening and other steps used to reduce the computational complexity do not significantly affect the overall parallel performance of the algorithm. In particular, the integral transformation still exhibits high parallel performance even though most of the screening is performed after the computational tasks have been distributed.

From the results obtained in this chapter, we can draw a few conclusions regarding efficient parallel implementation of local correlation methods: It may be possible to use parallel integral transformation algorithms developed for conventional correlation methods, with the appropriate modifications for local correlation. The parallel LMP2 integral transformation investigated here, which uses a modified version of an algorithm developed for canonical MP2 theory, achieves high parallel performance, and the algorithm displays good load balance also for large process counts even though the tasks, in this case shell pairs, are distributed before most of the integral screening is carried out. While distribution of data is also required in local correlation methods to avoid storage bottlenecks, it is possible to simplify the parallelization by replicating certain data structures that may be too large to replicate in a conventional correlation method, for instance, the double-substitution amplitudes. This replication of data, however, may require additional collective communication operations (such as an all-to-all broadcast of the double-substitution amplitudes in every iteration), and distribution of data should therefore also be considered. Finally, the sparsity of the data structures involved in local correlation methods necessitates an efficient way to handle distributed sparse multidimensional arrays.

References

1. Saebø, S., and P. Pulay. Local treatment of electron correlation. *Ann. Rev. Phys. Chem.* 44:213–236, 1993.
2. Schütz, M., G. Hetzer, and H.-J. Werner. Low-order scaling local electron correlation methods. I. Linear scaling local MP2. *J. Chem. Phys.* 111:5691–5705, 1999.
3. Pulay, P., and S. Saebø. Orbital-invariant formulation and second-order gradient evaluation in Møller-Plesset perturbation theory. *Theor. Chim. Acta* 69:357–368, 1986.
4. Hampel, C., and H.-J. Werner. Local treatment of electron correlation in coupled cluster theory. *J. Chem. Phys.* 104:6286–6297, 1996.
5. Nielsen, I. M. B., and C. L. Janssen. Local Møller-Plesset perturbation theory: A massively parallel algorithm. *J. Chem. Theor. Comp.* 3:71–79, 2007.

6. Wong, A. T., R. J. Harrison, and A. P. Rendell. Parallel direct four-index transformations. *Theor. Chim. Acta* 93:317–331, 1996.
7. Thakur, R., R. Rabenseifner, and W. Gropp. Optimization of collective communication operations in MPICH. *Int. J. High Perform. C.* 19:49–66, 2005.
8. A Linux® cluster consisting of nodes with two single-core 3.6 GHz Intel® Xeon® processors (each with 2 MiB of L2 cache) connected with a 4x Single Data Rate InfiniBand™ network using Mellanox Technologies MT25208 InfiniHost™ III Ex host adaptors and a Topspin 540 switch. The InfiniBand host adaptors were resident in a PCI Express 4x slot, reducing actual performance somewhat compared to using them in an 8x slot. MPICH2 1.0.5p4 was used with TCP/IP over InfiniBand IPoIB because this configuration provided the thread-safety required for some applications.

Appendices

A Brief Introduction to MPI

The Message-Passing Interface (MPI) is a software library providing the functionality for passing messages between processes in a distributed memory programming model. The MPI standard was developed to provide programmers with a single interface for message-passing that is portable to multiple architectures, and it is supported by essentially all distributed memory parallel computers used for scientific computing. We here give a brief introduction to MPI, demonstrating how to write a simple MPI program and discussing a few of the most widely used MPI functions for point-to-point and collective communication. This introduction is intended only to provide a rudimentary knowledge of MPI programming in a single-threaded context,* and the reader is referred elsewhere[1–4] for more comprehensive discussions. While the MPI standard[5] provides bindings for MPI functions for both ANSI C, C++, and Fortran, we will here use the C bindings only.

A.1 Creating a Simple MPI Program

In Figure A.1 we show a simple MPI program, written in C, illustrating several basic MPI calls used in most MPI programs to manage the MPI execution environment. The program includes the header file mpi.h

```
#include <mpi.h>
```

which contains all MPI type, function, and macro declarations and must be included to use MPI. The first MPI call in the program is `MPI_Init`, which initializes the MPI execution environment. `MPI_Init` has the signature

```
int MPI_Init(&argc, &argv)
```

and takes the same arguments, argc and argv, as those provided to the C main function. This initialization is always required; `MPI_Init` must be called by

* The use of MPI in a multi-threaded environment is briefly discussed in section 4.4.

```
#include <mpi.h>
#include <stdio.h>

int main(int argc, char **argv)
{
  int me;                                          /* process ID */
  int p;                                  /* number of processes */
  double t_start, t_end;

  MPI_Init(&argc, &argv);             /* initialize MPI environment */

  MPI_Comm_size(MPI_COMM_WORLD, &p);      /* get number of processes */
  MPI_Comm_rank(MPI_COMM_WORLD, &me);          /* assign process ID */

  MPI_Barrier(MPI_COMM_WORLD);          /* synchronize all processes */
  t_start = MPI_Wtime();                       /* record start time */

  /* ... Perform some work ... */

  MPI_Barrier(MPI_COMM_WORLD);          /* synchronize all processes */
  t_end = MPI_Wtime();                           /* record end time */

  if (me == 0) {
    printf("Wall time = %12.9f sec\n", t_end - t_start);
    printf("Number of processes = %5d \n", p);
  }

  MPI_Finalize();                     /* terminate MPI environment */
  return 0;
}
```

FIGURE A.1
An MPI program, written in C, illustrating several basic MPI operations. See text for details.

all processes, and it generally must precede any other MPI call.[†] The subsequent calls to `MPI_Comm_size` and `MPI_Comm_rank` (cf. Figure A.1) determine the number of processes and the rank of the calling process, respectively. The function declarations for these functions are

```
int MPI_Comm_size(MPI_Comm comm, int *size)
int MPI_Comm_rank(MPI_Comm comm, int *rank).
```

The first argument is an MPI communicator (of type `MPI_Comm`). An MPI communicator is an object associated with a group of processes that can communicate with each other, and it defines the environment for performing communication within this group. The default communicator is `MPI_COMM_WORLD`, which encompasses all processes. The `MPI_Comm_size` function determines the size of the associated communicator, that is, the number of processes

[†] The only MPI call that may be placed before `MPI_Init` is `MPI_Initialized`, which checks whether MPI has been initialized.

therein. MPI_Comm_rank determines the rank of the calling process within the communicator; the rank is a process identifier assigned to each process, and the processes within a communicator are numbered consecutively, starting at 0. The final MPI call in Figure A.1 is MPI_Finalize, which is required to terminate the MPI environment and must be called by all processes. The MPI_Finalize function, which has the signature

```
int MPI_Finalize()
```

must be the last MPI call in a program, and any MPI call placed after MPI_Finalize will produce an error.

All MPI functions (except timing calls) have an integer return type, and the integer returned is an error value, which can be used to determine whether the function completed successfully. Upon successful completion of a function call, the return value will be equal to MPI_SUCCESS, and a code segment like the following

```
rc = MPI_Init(&argc,&argv);
if (rc != MPI_SUCCESS} {
  /* ... abort execution ... */
}
```

can thus be used to check whether MPI was successfully initialized.

The program in Figure A.1 also illustrates the use of MPI timing calls via the MPI_Wtime function

```
double MPI_Wtime()
```

which returns the wall clock time (in seconds) on the calling process measured relative to some point in the past that is guaranteed to remain the same during the lifetime of the process. The values returned by MPI_Wtime by different processes, however, may not be synchronized, and elapsed times should therefore be computed from timing calls made by the same process. In Figure A.1, the elapsed time for a segment of the program is measured by placing calls to MPI_Wtime before and after this segment, and the timings, as measured on process 0, are printed out. To make sure that the longest time required by any process is recorded, all processes are synchronized before making calls to the MPI_Wtime function; the synchronization is performed with the MPI_Barrier function (see section A.2.2).

A.2 Message-Passing with MPI

Having introduced the basic MPI calls for managing the execution environment, let us now discuss how to use MPI for message-passing. MPI provides support for both point-to-point and collective communication operations.

TABLE A.1

Types of arguments required by various MPI functions

Buffer:	Buffer (of type void*) holding the message to be sent or received
Count:	Integer specifying the number of elements in the message buffer
Destination:	Integer specifying the rank of the process to receive the message
Source:	Integer specifying the rank of the process sending the message
Tag:	Tag (nonnegative integer) identifying the message
Data type:	Specification (of type `MPI_Datatype`) of the type of data to be sent or received (e.g., `MPI_INT`, `MPI_DOUBLE`, `MPI_CHAR`)
Root:	Integer specifying the rank for the root process
Status:	Object (of type `MPI_Status`) containing information about a received message
Request:	Handle (of type `MPI_Request`) used for non-blocking send and receive to make it possible to determine if the operation has completed
Operation:	Operation (of type `MPI_Op`) to be performed in an MPI reduction (e.g., `MPI_MAX`, `MPI_MIN`, `MPI_SUM`, `MPI_PROD`)
Flag:	Integer returned as either logical true (1) or false (0)
Communicator:	Object (of type `MPI_Comm`) representing a group of processes that can communicate with each other. `MPI_COMM_WORLD` can be used to specify all processes

There are a very large number of such operations, and we will here briefly discuss some of the most commonly employed functions. A number of different argument types are required by these functions, and in Table A.1 we list all argument types required by MPI functions included in this appendix.

A.2.1 Point-to-Point Communication

We have seen in section 3.1 that point-to-point communication operations can be either blocking or non-blocking, and MPI provides support for a number of point-to-point operations of both kinds. A few of the most basic MPI point-to-point communication operations are listed in Table A.2. They include the blocking send and receive operations `MPI_Send` and `MPI_Recv` and their non-blocking immediate mode counterparts `MPI_Isend` and `MPI_Irecv`. The blocking wait operation `MPI_Wait` can be posted after a non-blocking send or receive operation to wait for the preceding operation to complete, and the non-blocking function `MPI_Test` tests whether a non-blocking send or receive operation has completed. Examples of the use of MPI blocking and non-blocking point-to-point operations can be found in Figures 3.2 and 5.2. A few blocking send operations other than `MPI_Send` are included in Table A.2, namely, `MPI_Ssend`, `MPI_Bsend`, and `MPI_Rsend`.

The `MPI_Send` function, in some implementations (for certain data sizes), is buffered so that the send buffer will immediately be copied into a local buffer used for the message transfer, and the send function in that case will return independently of whether a corresponding receive has been posted at the receiving end. The `MPI_Ssend` function, on the other hand, is guaranteed not to return until the message transfer has started, and, hence, requires a matching receive on the receiving process in order to complete. If a buffered

TABLE A.2

C bindings for some commonly used MPI point-to-point communication operations; output parameters are underlined. All of the operations have return type `int`. The non-blocking counterparts to `MPI_Ssend`, `MPI_Bsend`, and `MPI_Rsend` are `MPI_Issend`, `MPI_Ibsend`, and `MPI_Irsend`, respectively; these functions, which are not shown in the table, take the same arguments as `MPI_Isend`

Action	MPI Function

Blocking receive:	`MPI_Recv(void *buf, int count, MPI_Datatype datatype,`
	`int source, int tag, MPI_Comm comm,`
	`MPI_Status *status)`

Blocking send:	`MPI_Send(void *buf, int count, MPI_Datatype datatype,` `int dest, int tag, MPI_Comm comm)`
Blocking synchronous send:	`MPI_Ssend(void *buf, int count, MPI_Datatype datatype,` `int dest, int tag, MPI_Comm comm)`
Blocking buffered send:	`MPI_Bsend(void *buf, int count, MPI_Datatype datatype,` `int dest, int tag, MPI_Comm comm)`
Blocking ready send:	`MPI_Rsend(void *buf, int count, MPI_Datatype datatype,` `int dest, int tag, MPI_Comm comm)`

Non-blocking receive:	`MPI_Irecv(void *buf, int count, MPI_Datatype datatype,`
	`int source, int tag, MPI_Comm comm,`
	`MPI_Request *request)`

Non-blocking send:	`MPI_Isend(void *buf, int count, MPI_Datatype datatype,`
	`int dest, int tag, MPI_Comm comm,`
	`MPI_Request *request)`

Wait for non-blocking operation to complete:	`MPI_Wait(MPI_Request *request, MPI_Status *status)`
Test whether non-blocking operation has completed:	`MPI_Test(MPI_Request *request, int *flag,` `MPI_Status *status)`

send is desired, the function `MPI_Bsend` can be used, and this can sometimes improve performance by allowing faster return of control to the calling process. The ready send `MPI_Rsend` can be posted only after a matching receive has been posted by the receiving process (or the result will be undefined), and in this case using `MPI_Rsend` instead of `MPI_Send` may (depending on the MPI implementation) improve performance.

A.2.2 Collective Communication

A number of the most widely used collective communication operations provided by MPI are listed in Table A.3. The collective operations have been grouped into operations for data movement only (broadcast, scatter, and gather operations), operations that both move data and perform computation on data (reduce operations), and operations whose only function is to synchronize processes. In the one-to-all broadcast, `MPI_Bcast`, data is sent from one process (the root) to all other processes, while in the all-to-all broadcast, `MPI_Allgather`, data is sent from every process to every other process (one-to-all and all-to-all broadcast operations are discussed in more detail in section 3.2). The one-to-all scatter operation, `MPI_Scatter`, distributes data from the root process to all other processes (sending different data to different processes), and the all-to-one gather, `MPI_Gather`, is the reverse operation, gathering data from all processes onto the root.

TABLE A.3

C bindings for some widely used MPI collective operations; output parameters are underlined. All of the operations have return type `int`. For `MPI_Bcast`, the `buf` parameter is an output parameter on all processes except the root

Action	MPI Function
Data Movement Only	
One-to-all broadcast:	`MPI_Bcast(void *buf, int count, MPI_Datatype datatype,` `int root, MPI_Comm comm)`

All-to-all broadcast:	`MPI_Allgather(void *sendbuf, int sendcount,`
	`MPI_Datatype sendtype, void *recvbuf,`
	`int recvcount, MPI_Datatype recvtype,`
	`MPI_Comm comm)`

One-to-all scatter:	`MPI_Scatter(void *sendbuf, int sendcount,`
	`MPI_Datatype sendtype, void *recvbuf,`
	`int recvcount, MPI_Datatype recvtype,`
	`int root, MPI_Comm comm)`

All-to-one gather:	`MPI_Gather(void *sendbuf, int sendcount,`
	`MPI_Datatype sendtype, void *recvbuf,`
	`int recvcount, MPI_Datatype recvtype`
	`int root, MPI_Comm comm)`

Data Movement and Computation	

All-to-one reduction:	`MPI_Reduce(void *sendbuf, void *recvbuf, int count,`
	`MPI_Datatype datatype, MPI_Op op,`
	`int root, MPI_Comm comm)`

All-reduce:	`MPI_Allreduce(void *sendbuf, void *recvbuf, int count,`
	`MPI_Datatype datatype,`
	`MPI_Op op, MPI_Comm comm)`

Synchronization Only	
Barrier synchronization:	`MPI_Barrier(MPI_Comm comm)`

The all-to-one reduction and all-reduce operations, `MPI_Reduce` and `MPI_Allreduce`, perform a reduction operation on data across all processes and place the result on either the root process (`MPI_Reduce`) or on all processes (`MPI_Allreduce`). The type of operation performed is specified via the `op` argument of type `MPI_Op`, and the types of operations that can be performed include, for example, finding the maximum or minimum value across a data set or performing a multiplication or addition of the numbers in the set, and user defined operations can be specified as well. The synchronization operation listed in Table A.3, `MPI_Barrier`, is a barrier synchronization: when encountering the barrier, a process cannot continue until all processes have reached the barrier, and, hence, all processes are synchronized.

To illustrate the use of collective communication operations, we show in Figure A.2 an MPI program that employs the collective communication operations `MPI_Scatter` and `MPI_Reduce`; the program distributes a matrix (initially located at the root process) across all processes, performs some computations on the local part of the matrix on each process, and performs a global summation of the data computed by each process, putting the result on the root process.

```
#include <mpi.h>
#include <stdio.h>
#define ndata 10

int main (int argc, char *argv[])
{
  int me, p; /* process ID and number of processes */
  double local_result, result;
  double A[ndata*ndata];
  double my_row[ndata];

  MPI_Init(&argc, &argv);
  MPI_Comm_rank(MPI_COMM_WORLD, &me);
  MPI_Comm_size(MPI_COMM_WORLD, &p);
  /* quit if number of processes not equal to ndata */
  if (p != ndata) {
    if (me == 0) printf("Number of processes must be %d. "
                        "Quitting. \n", ndata);
    abort();
    }

  /* If root process, read in A matrix from somewhere */
  if (me == 0) read_data(A);

  /* Scatter rows of A across processes (one row per process) */
  MPI_Scatter(A, ndata, MPI_DOUBLE, my_row, ndata,
              MPI_DOUBLE, 0, MPI_COMM_WORLD);

  /* Do some computation on the local row of A */
  local_result = compute(my_row);

  /* Add local_result from all processes and put on root process */
  MPI_Reduce(local_result, result, 1, MPI_DOUBLE, MPI_SUM, 0,
             MPI_COMM_WORLD);

  /* If root process, print out result */
  if (me == 0) printf("Result: %lf \n", result);

  MPI_Finalize();
  return 0;
}
```

FIGURE A.2
An MPI program, written in C, illustrating the use of the MPI collective operations `MPI_Scatter` and `MPI_Reduce`. The root process (process 0) scatters rows of a matrix across all processes; each process then performs some computations on its own row and, finally, the sum of the results from all processes is put on the root process.

References

1. Gropp, W., E. Lusk, and A. Skjellum. *Using MPI*, 2nd edition. Cambridge: MIT Press, 1999.
2. Gropp, W., E. Lusk, and R. Thakur. *Using MPI-2*. Cambridge: MIT Press, 1999.
3. Quinn, M. J. *Parallel Programming in C with MPI and OpenMP*. New York: McGraw-Hill, 2003.
4. Pacheco, P. S. *Parallel Programming with MPI*. San Francisco: Morgan Kaufmann, 1997.
5. The MPI-2 standard for the Message-Passing Interface (MPI) is available on the World Wide Web at http://www.mpi-forum.org.

B

Pthreads: Explicit Use of Threads

Pthreads, also known as POSIX® Threads, as specified by the IEEE POSIX 1003.1c standard, provides fairly portable support for multi-threading. It uses a library interface that enables the user to explicitly manipulate threads, for instance, to start, synchronize, and terminate threads. The programmer fully controls when threads are created and what work they perform, and this provides great flexibility, albeit at the cost of greater complexity than the OpenMP approach discussed in Appendix C. In the following we will give a brief introduction to programming with Pthreads; for more complete information, good guides to multi-threaded programming and Pthreads can be found elsewhere.[1,2]

We will illustrate programming with Pthreads using the simple example shown in Figure B.1 in which a number of tasks are distributed between two threads. To use Pthreads, it is necessary to include the Pthreads header file `pthread.h`, which declares the function signatures for the Pthreads library along with the required data types. A few of the most commonly used Pthreads functions are listed in Table B.1, and their arguments are described in Table B.2. A key feature required by a Pthreads program is a start routine, which is the function that will be run by the threads. In our example, the start routine for the created threads is the `worker` function. Having defined a start routine, threads can be created by means of the `pthread_create` function call, which takes the start routine as one of its arguments. In our example program, the threads running the `worker` routine obtain their tasks from a shared pointer to a variable, which is named `task_iter` and is of type `task_iterator_t`. The `task_iter` variable contains an integer (`next_task`) that gives the next unassigned work unit, an integer that gives the last work unit (`max_task`), and a mutual exclusion lock (`mutex`). Because all start routines must have the same signature, arguments are passed using a pointer to `void`, and the start routine must cast this into whatever type is needed. In our example, the `task_iter` variable is the argument for the start routine.

Let us take a closer look at the creation of the threads. The `pthread_create` function, which creates the threads, requires a few arguments in addition to the start routine mentioned above (see Table B.1). One of these

```
#include <pthread.h>
typedef struct {
  volatile int next_task;
  int max_task;
  pthread_mutex_t mutex;
} task_iterator_t;

void *worker(void *arg) {
  task_iterator_t *task_iter = (task_iterator_t*)arg;
  int my_task;
  while (1) {
    pthread_mutex_lock(&task_iter->mutex);
    my_task = task_iter->next_task++;
    pthread_mutex_unlock(&task_iter->mutex);
    if (my_task > task_iter->max_task) return 0;
    /* ... process work unit 'my_task' ... */
  }
}

int main() {
  /* Declarations. */
  pthread_attr_t attr;
  task_iterator_t task_iter;
  pthread_t thread_1, thread_2;

  /* Initialize. */
  pthread_attr_init(&attr);
  pthread_attr_setstacksize(&attr,8000000);
  task_iter.max_task = 100;
  task_iter.next_task = 0;
  pthread_mutex_init(&task_iter.mutex,NULL);

  /* Start threads. */
  pthread_create(&thread_1,&attr,worker,&task_iter);
  pthread_create(&thread_2,&attr,worker,&task_iter);

  /* Wait for threads to complete. */
  pthread_join(thread_1,NULL);
  pthread_join(thread_2,NULL);

  /* Clean-up and return. */
  pthread_attr_destroy(&attr);
  pthread_mutex_destroy(&task_iter.mutex);
  return 0;
}
```

FIGURE B.1
An example of distribution of work among threads using the Pthreads library.

TABLE B.1

Several of the most common Pthreads functions. All of these functions return a zero to indicate success, with the exception of pthread_exit, which returns void. See Table B.2 for a description of the arguments

Action	Pthreads Function	Arguments
Create a thread:	pthread_create	pthread_t *thread const pthread_attr_t *attr void *(*start_routine)(void*) void *arg
Initialize attributes:	pthread_attr_init	pthread_attr_t *attr
Release attributes:	pthread_attr_destroy	pthread_attr_t *attr
Initialize mutex attributes:	pthread_mutexattr_init	pthread_mutexattr_t *mattr
Release mutex attributes:	pthread_mutexattr_destroy	pthread_mutexattr_t *mattr
Create a mutex:	pthread_mutex_init	pthread_mutex_t *mutex const pthread_mutexattr_t *mattr
Destroy a mutex:	pthread_mutex_destroy	pthread_mutex_t *mutex
Lock a mutex:	pthread_mutex_lock	pthread_mutex_t *mutex
Unlock a mutex:	pthread_mutex_unlock	pthread_mutex_t *mutex
Terminate calling thread:	pthread_exit	void *value_ptr
Wait for a thread:	pthread_join	pthread_t thread void **value_ptr

arguments is the argument required by the start routine, namely, in our example, task_iter. Another required argument for pthread_create is a pthread_t pointer that will be used to store an identifier for the thread. Additionally, a pointer to an attribute object, pthread_attr_t, specifying the properties of the thread, and possibly hints regarding how to run it, is needed (a value of NULL can be provided to use the default attributes). In our example, this attribute argument is named attr; it has been initialized by calling the pthread_attr_init function, and the stack-size attribute has been set with the pthread_attr_setstacksize function.

In the example in Figure B.1, the pthread_create function is called twice, creating two new threads in addition to the main thread; the two new threads each start running the worker routine, and the main thread keeps running, so three threads are active at the same time. The main thread can wait for the execution of the two other threads to complete and ignore their return

TABLE B.2

Types of arguments required by various Pthreads functions

thread	Type pthread_t identifier for a thread.
attr	Type pthread_attr_t thread attribute descriptor; attributes include properties such as the thread's stacksize and scheduling policy.
start_routine	Pointer to a function that takes and returns a pointer to void. This routine will be run in a separate thread.
arg	A pointer to void used as an argument to a start_routine.
value_ptr	A pointer to void* used to place the result returned by a thread's start_routine.
mutex	A mutex object of type pthread_mutex_t.
mattr	Type pthread_mutexattr_t mutex attribute descriptor; mutex attributes can specify that the mutex can be locked recursively, for example.

values by calling the `pthread_join` function for each of the two threads as shown in the figure. Before performing the join, the main thread could have run the `worker` routine as well for a total of three threads concurrently running the routine. Finally, the attribute object and the mutex can be destroyed by calling the `pthread_attr_destroy` and `pthread_mutex_destroy` functions.

The mutex in our program example is required to ensure that only one thread at a time reads and updates the shared work variable, `task_iter` (without the mutex, simultaneous updates of `task_iter` could occur, namely, two threads updating `task_iter->next_task` at the same time). The mutex must be initialized with a call to `pthread_mutex_init`. Here, the default mutex attributes are satisfactory, so a `NULL` is passed as the mutex attribute argument. The `pthread_mutex_lock` call in the `worker` function acquires the mutex lock, and only one thread at a time can hold the lock on the mutex. When one thread holds the lock, the other thread will block at the call to `pthread_mutex_lock` until `pthread_mutex_unlock` is called by the holding thread. At that time, the thread blocked by `pthread_mutex_lock` will be allowed to proceed.

We have not yet addressed the initialization of the variable `task_iter`, which is passed as an argument to the `worker` function and specifies information about the tasks to be processed. A single datum of type `task_iterator_t` needs to be allocated and initialized. In the example in Figure B.1, the field `next_task`, storing the number of the next work unit, is initialized to zero, and the `max_task` field is initialized to the largest allowed value for `next_task`; the field `mutex` is initialized by calling `pthread_mutex_init` as explained above. Finally, note that in the program in Figure B.1, we have used the `volatile` type modifier to inform the compiler that the `next_task` element of the `task_iter_t` structure could be changed outside the context of the current thread. Just before a volatile datum is used, it is read from memory, and immediately after it is modified, it is written to memory. As used here, the `volatile` modifier is not strictly necessary because the possibility of side effects in the function calls just before and after the use of the `next_task` variable force the compiler to perform the read and write anyway. Nonetheless, in other contexts `volatile` may be required, and it is also good style to use it to remind the programmer of what data could be changed by other threads.

Pthreads provides a large amount of flexibility as well as complete control over how threads are used; additionally, Pthreads provides *condition variables* that permit development of complex locking protocols that wait for a condition to be satisfied. In Appendix C we will discuss a simpler alternative to Pthreads, OpenMP. Section 4.3 gives another example of a Pthreads program and compares it to the equivalent OpenMP and message-passing programs.

References

1. Lewis, B., and D. J. Berg. *Threads Primer: A Guide to Multithreaded Programming*. Mountain View, CA: SunSoft Press, 1996.
2. Lewis, B., and D. J. Berg. *Multithreaded Programming with PThreads*. Upper Saddle River, NJ: Prentice Hall, 1997.

C

OpenMP: Compiler Extensions for Multi-Threading

The Pthreads standard discussed in Appendix B is very flexible and provides a large amount of functionality, but parallel programming with Pthreads can be complicated and requires implementation of code written specifically with multi-threading in mind. OpenMP provides an alternative, and somewhat simpler, way to use multi-threading in parallel applications. OpenMP is an extension to the Fortran, C, and C++ languages designed to make it possible to easily take advantage of multi-threading in both new programs and existing serial code. OpenMP also provides several work-sharing constructs that are much more high level than those provided by Pthreads. OpenMP is used by providing the compiler with code annotations. These annotations are ignored by compilers that do not support OpenMP, making it possible to maintain backwards compatibility. OpenMP has been available in commercial compilers for some time, and it is also available in recent versions of the GNU Compiler Collection (GCC),[1] making its adoption rate likely to increase.

Despite the relative ease of using OpenMP, the issues discussed in section 4.1 pertaining to threads modifying resources shared by multiple threads also apply to OpenMP programming. It is left to the programmer to ensure that the necessary protections are applied to these resources, and multi-threaded programming with OpenMP can be particularly hazardous for programmers modifying preexisting code that was not written with multi-threading in mind. In the following we give a brief introduction to OpenMP, illustrating its range of functionality and how to write small programs. Complete details for programming with OpenMP, along with numerous examples, can be found in the OpenMP specification.[2]

We will use C code fragments to illustrate OpenMP; the equivalent C++ code fragments would be similar to those for C, but the use of OpenMP in Fortran looks somewhat different, although the concepts are the same. The C and C++ code annotations in OpenMP are specified via the pragma mechanism, which provides a way to specify language extensions in a portable way.

TABLE C.1

A few of the primary OpenMP functions. These functions are declared in the `omp.h` system include file

OpenMP Function	Arguments	Action
`omp_get_num_threads`		Returns the number of threads in the current team.
`omp_get_thread_num`		Returns the number of the current thread.
`omp_set_num_threads`	`int` *num_threads*	Sets the number of threads to use.
`omp_get_num_procs`		Returns the number of processors.
`omp_get_nested`		Returns true if nested parallelism is enabled.
`omp_set_nested`	`int` *nested*	Enables nested parallelism if and only if *nested* is true and the implemenation supports it.
`omp_get_wtime`		Returns a double containing the wall time in seconds starting from an arbitrary reference time.
`omp_get_wtick`		Returns a double giving the clock resolution in seconds.

Pragmas take the form of a C preprocessor statement,* and compilers that do not recognize a pragma will ignore them.† OpenMP pragmas begin with the keyword `omp`, which is followed by the desired directive:

```
#pragma omp directive
```

The directive will apply to the immediately following statement only. If the directive needs to apply to several statements, the statements can be grouped together by enclosing them between a `{` and a `}`. In the following, the term "block" will be used both for single statements and for multiple statements grouped together in this way.

C.1 Functions

OpenMP provides functions to query the OpenMP environment, control the behavior of OpenMP, acquire and release mutexes, and obtain timing information. A few of these functions are summarized in Table C.1. The `omp.h`

---

* C99 permits pragmas also to be given with an operator notation.

† This calls for the use of some caution when using OpenMP. In many cases, pragmas that are mistyped will not be used or reported—they are just ignored.

system include file, which declares these functions, must be included as follows if the OpenMP functions are to be used:

```
#include <omp.h>
```

OpenMP also provides a few environment variables; one of these is the `OMP_NUM_THREADS` variable, and a Bash shell statement of the form

```
export OMP_NUM_THREADS=k
```

where k is a positive integer sets the number of threads to be used equal to k.

C.2 Directives

The OpenMP directives shown in Table C.2 can be divided into those that specify parallel regions, work-sharing constructs, and synchronization constructs. The primary OpenMP directive is `parallel`, and the block of code that follows this statement is a parallel region that will be executed by multiple threads, namely a team of threads created by the thread encountering

TABLE C.2

Selected OpenMP directives and their effect. C/C++ directives are shown, and similar directives exist for Fortran

Parallelism Directives	
`parallel`	The key OpenMP directive. Forms a team of threads that all execute the following program block.
Work-Sharing Directives	
`for`	Distribute the work in the following `for` statement among all the threads in the current team.
`sections`	Make the current thread team execute in parallel the `section` constructs in the following program block.
`single`	Execute the following program block on any single thread in the current team.
Synchronization Directives	
`critical`	Serialize execution of the following program block with respect to all threads in all teams.
`master`	Execute the next program block only on the current team's master thread (thread number 0).
`flush`	Force data in memory to be consistent with any potential temporary copies.
`barrier`	No thread in the current team passes the barrier until all reach it.
`atomic`	The storage location modified in the following statement is atomically modified with respect to all threads in all teams.
`ordered`	Execute the following program block in order as if loop iterations were performed serially.

the parallel directive and consisting of this thread and zero or more additional threads. The default number of threads on which to run is defined by the OpenMP implementation, and the default can be overridden by setting the OMP_NUM_THREADS environment variable or by calling omp_set_num_threads. It is also possible to give clauses to the parallel pragma that affect the number of threads that will be created, although these will not be discussed here. Furthermore, if the parallel region is nested inside of another, then additional settings and implementation details will affect the number of additional threads created. In the following example, a team of threads is created, and each of these threads executes the block of code following the pragma:

```
  double x[100];
#pragma omp parallel
  {
    int nthread = omp_get_num_threads();
    int mythread = omp_get_thread_num();
    for (i=mythread; i<100; i+=nthread) {
      x[i] = process_iterand(i);
    }
  }
```

Here, process_iterand is a thread-safe routine that produces elements of the x vector, and the programmer has explicitly distributed the work among the threads in a round-robin fashion.

C.2.1 Work-Sharing Constructs

OpenMP provides work-sharing constructs that make it unnecessary for the programmer to provide code that explicitly distributes the work among threads. A work-sharing construct binds to the thread team created by the innermost parallel directive within which the construct is nested.

C.2.1.1 The for Directive

The following code fragment illustrates the use of the for directive:

```
  double x[100];
#pragma omp parallel
  {
    int i;
#pragma omp for
    for (i=0; i<100; i++) {
      x[i] = process_iterand(i);
    }
  }
```

The `for` directive binds to the innermost parallel region in which it is nested. It distributes the work in the following `for` loop among the threads in the team established by the corresponding `parallel` directive. In this example, the `process_iterand` function is called 100 times with all integers from 0 to 99, inclusive, and it could be called concurrently from multiple threads with any order of the iterands. The `for` loop is required to have a special structure: the counter variable must be a signed integer type and it must not be modified except in the `for` statement's increment expression; the completion test must use one of the >=, <=, <, or > relational operators; and the loop variable must be initialized to, compared to, and incremented or decremented by integer expressions that are not modified within the loop. Options to control how the work is scheduled can be given to the `for` directive or in an environment variable.

Often, a scalar (that is, a single variable in a certain memory location) is modified in each pass through a `for` loop. Contributions to the scalar must be handled in a special way since all threads must update its value using proper locking protocols. This can be accomplished by adding a `reduction` clause to the line containing a `parallel` or work-sharing directive. For instance, a dot product routine could be parallelized as:

```
  double r;
#pragma omp parallel
#pragma omp for reduction(+: r)
  for (i=0; i<n; i++)
    r += x[i]*x[i];
```

This `reduction` clause notifies the compiler that the `r` variable is to be reduced by applying the + operator; a private copy of `r` will be created for each thread to use during the execution of the loop, and at the end, the reduction operation is then applied on these local copies, in this case adding the contributions from all threads. Other reduction operators include `*`, `-`, `&`, `^`, `|`, `&&`, and `||`. A comma-separated list of variables can be given, and multiple reduction clauses with different operators can be specified. The order in which the reduction is performed is not defined, so numerical round-off error for floating point reductions can result in slightly different results for multiple runs with the same input data. The consequences of omitting the `reduction` clause in the above example are dire: the program will compile without warnings, and regression tests can produce the correct result, but actual production runs could produce an incorrect, nondeterministic result. A bug of this type could be very difficult to find.

The accumulation of `x[i]*x[i]` into `r` in the above example is a critical section (see section 4.1) because `r` is shared among all threads. The (almost) associative and commutative properties of the addition and other permitted operators allow the `reduction` operator to provide better performance than would be possible by obtaining a mutual exclusion lock for each accumulation into `r`.

C.2.1.2 The `sections` Directive

The `sections` directive specifies another work-sharing construct. The program block following a `sections` directive contains a series of nested blocks, and each of the blocks, except for the first (for which it is optional) must be preceded by a `section` directive, as shown below:

```
  double a, b;
#pragma omp parallel
#pragma omp sections
  {
#pragma omp section
    {
      a = compute_a();
    }
#pragma omp section
    {
      b = compute_b();
    }
  }
```

The execution of these blocks will be divided among the threads in the team, and each `section` block will be executed exactly once by a thread belonging to the current team created by the innermost containing parallel region.

C.2.1.3 Compound Directives

As can be seen from the above examples, a `parallel` directive is often followed immediately by a work-sharing directive. Therefore, OpenMP supports combined directives where the work-sharing directive is on the same line as the `parallel` directive, as follows:

```
  double x[100];
  int i;
#pragma omp parallel for
  for (i=0; i<100; i++) {
    x[i] = process_iterand(i);
  }
```

which has the same effect as the first example of the `for` directive in section C.2.1.1.

C.2.2 Synchronization Constructs

During the multi-threaded parallel execution inside of an OpenMP work-sharing region, it is sometimes necessary to explicitly coordinate the activity of the threads. We have already seen an example of implicit synchronization with the `reduction` clause of the `for` directive, which causes each thread to ensure that updates to the reduction variables occur in a thread-safe manner.

There is also an implicit synchronization occurring after a thread completes a work-sharing construct—threads wait for all the others to finish unless a `nowait` clause is specified. Other situations requiring synchronization can be dealt with by using one of the following explicit synchronization directives.

C.2.2.1 The `critical` Directive

If a non-thread-safe function is called inside of a `for` loop, it is necessary to ensure that the function call is made by only one thread at a time. Such a critical section can be protected by a `critical` directive, which serializes the execution of the program block following it. In the following example, the progress of each thread is reported using `printf`

```
  int i;
#pragma parallel for
  for (i=0; i<10; i++) {
    double r = compute_result(i);
    int mythread = omp_get_thread_num();
#pragma omp critical
    printf("in thread %2d, r = %12.8f\n",
           mythread, r);
}
```

The `printf` function modifies the output stream, which is shared among all threads, and depending on the details of the system C library, `printf` may or may not be thread-safe. By preceding the call to `printf` with a `critical` directive, `stdout` can be modified by only one thread at a time, ensuring that `printf` will work correctly.

C.2.2.2 The `atomic` Directive

The `critical` directive imposes a performance penalty by forcing complete serialization among all threads. An alternative synchronization directive, the `atomic` directive, can be used to prevent a memory location from being modified by more than one thread at a time without requiring all threads to be serialized. Consider, for example, the formation of a histogram by the aggregation of data into a few bins. Because different threads may have to update data in the same bin, the update is a critical section. However, the `critical` directive would result in complete serialization of all threads as they sum their contributions into the array. In this case, only threads that try to add contributions to the same bin need to be serialized. The `atomic` directive ensures that the memory location being modified in the statement following it is read, modified, and written out again atomically. As a result, each thread will always have a consistent view of that memory. In the following example, a histogram is formed from the results of the `compute_value` function:

```
int i;
int histogram[10];
for (i=0; i<10; i++) histogram[i] = 0;
```

```
#pragma omp parallel for
  for (i=0; i<10000; i++) {
    int index = (int)(compute_value(i)*10.0);
    if (index < 0) index = 0;
    if (index >= 10) index = 9;
#pragma omp atomic
    histogram[index]++;
  }
```

C.2.2.3 The `flush` Directive

The `flush` directive is related to the `volatile` keyword in C/C++, although it is more flexible. A thread may create a temporary copy of a variable, for example, in a register, making it possible for different threads to have different values for a shared variable at the same time. Memory consistency is enforced when all threads needing a consistent copy of a variable encounter the `flush` directive. Specific variables can be named with the `flush` directive, or the variable names can be left out. In the latter case, all data visible to the thread is flushed. Below, the memory for `x` and `y` is flushed:

```
#pragma parallel {
   /* ... */
#pragma omp flush(x,y)
   /* ... */
}
```

Flushes are automatically issued in several situations: at `barrier` directives, and before and after regions affected by `parallel`, `critical`, or `ordered` directives. Also, when the `nowait` clause is not given to a `for` or `sections` directive, a flush will be implicit upon exit from the associated region. Flushes are also implicitly performed during explicit lock manipulation operations. The data updated following the `atomic` directive is flushed before and after the atomic region is executed. Because of these automatic flushes, it is seldom necessary for the programmer to explicitly flush data.

C.2.2.4 The `master` Directive

This directive specifies that the following block is only executed on the master thread within a thread team. The following code fragment, therefore, will print only one line of output:

```
#pragma omp parallel
  {
#pragma omp master
    printf("region entered by master thread\n");
  }
```

C.2.2.5 The `ordered` Directive

The `ordered` directive causes the block following it to be processed sequentially in the same order as the loop iterations. Note that only the part of the parallel block that is outside of the scope of the `ordered` directive will be performed in parallel, and using the `ordered` directive can therefore cause severely reduced performance. In the following example, the `ordered` directive is used to handle a recurrence relation requiring that x_{i-1} be known to compute x_i. Note that the `for` in which `ordered` is nested must contain an `ordered` clause.

```
  int i;
  double x[100];
  x[0] = 0.0;
#pragma omp parallel for ordered
  for (i=1; i<100; i++) {
    double delta_x = compute_delta_x(i);
#pragma omp ordered
    x[i] = x[i-1] + a;
  }
```

C.2.3 Data-Sharing Directive Clauses

Above we have discussed various OpenMP directives as well as a couple of directive clauses (`reduction` and `nowait`). In OpenMP there are a number of directive clauses that can be used to specify the sharing status of variables within a parallel region, and we will discuss a few of these in the following. When an OpenMP program is executing within a thread-parallel region, each thread can read from and write to variables that were defined in the scope surrounding the parallel region as well as new variables declared within the parallel region. It is essential for the programmer to understand whether referenced variables are private to each thread or shared by all of the threads concurrently executing that region. The sharing status of a variable can either be predetermined by the type qualifiers and scope for that variable or implicitly determined. When the sharing status is implicitly determined, the programmer can alter it by using a clause that specifically defines the sharing status. The OpenMP specification gives detailed rules for what is shared and what is private. For the most part, variables declared in the scope outside of the parallel region of interest are by default shared. Variables allocated with automatic storage within the parallel region are always private. However, static variables are by default shared, and data allocated on the heap (with `malloc` or `new`) are always shared. Also, the iterand in the `for` statement following a `for` directive is always private.

C.2.3.1 *The `private` Clause*

Variables that implicitly have shared status can be made private by explicitly naming them in a `private` clause given to a `parallel` directive. Such private variables are given independent storage locations in each thread in the current thread team. The initial and final values of private variables are not defined. However, if the variable is named in a `firstprivate` clause, instead of a `private` clause, the variable will be private and each thread's private copy of the variable will automatically be initialized to its value at the point the parallel construct is encountered. If the variable must have a well-defined value after execution of the parallel region has completed, the `lastprivate` clause will set the variable to its value computed in the last iteration of a `for` construct or the last `section` region of a `sections` construct.

C.2.3.2 The `shared` Clause

The `shared` clause explicitly specifies that the named variables will be shared among all threads. This is particularly useful when the `default(none)` clause is used (see the next section), in which case all shared variables declared in an outer scope must be explicitly named in a `shared` directive.

C.2.3.3 The `default` Clause

The `default` clause takes one argument that gives the default scope for those variables in the parallel construct that have implicit sharing status. The possible arguments in C/C++ are `none` and `shared`; specifying `none` will make it illegal to reference any variable with implicit sharing status declared in an outer scope, while specifying `shared` will cause these variables to be shared by default. Using `default(none)` provides an extra degree of safety by preventing the programmer from accidentally assuming that a shared variable is private, as in the following example:

```
  int i,j;
  double x[10][10];
#pragma omp parallel for shared(x) default(none)
  for (i=0; i<10; i++) {
    for (j=0; j<10; j++) {
      x[i][j] = compute_x(i,j);
    }
  }
```

A compiler supporting OpenMP will reject this code: the `default(none)` clause is given, but the sharing status of `j` is implicit. A segment of code like the above, however, could easily lead to a subtle error that would be difficult to detect; if the `default(none)` clause had accidentally been omitted in the code example above, the program would have compiled without warnings, and `j` would be shared among all of the threads, possibly causing illegal memory overwrites or causing some elements of `x` to not be computed. The code listed above will compile and work as intended, that is, `j` will be private, if the `private(j)` clause is also given with the `parallel` directive.

References

1. The GNU Compiler Collection. http://gcc.gnu.org.
2. OpenMP Application Program Interface, Version 2.5. May 2005. Available from http://www.openmp.org.

Index

D

E

F

G

H

I

K

L

M

N

O

P

Q

R

S

Milton Keynes UK
Ingram Content Group UK Ltd.
UKHW040101071024
449327UK00019B/732